Work with ionising radia

Ionising Radiations Regulations 2017

Approved Code of Practice and guidance

London: TSO

HSE

tso
a Williams Lea company

Published by TSO (The Stationery Office), part of Williams Lea,
and available from:

Online
https://books.hse.gov.uk/

Mail, Telephone, Fax & E-mail
TSO
PO Box 29, Norwich, NR3 1GN
Telephone orders/General enquiries: 0333 202 5070
Fax orders: 0333 202 5080
E-mail: customer.services@tso.co.uk
Textphone 0333 202 5077

TSO@Blackwell and other Accredited Agents

Approved Code of Practice
This Code has been approved by the Health and Safety Executive, with the consent of the Secretary of State. It gives practical
advice on how to comply with the law. If you follow the advice you will be doing enough to comply with the law in respect of
those specific matters on which the Code gives advice. You may use alternative methods to those set out in the Code in order
to comply with the law.

However, the Code has a special legal status. If you are prosecuted for breach of health and safety law, and it is proved that
you did not follow the relevant provisions of the Code, you will need to show that you have complied with the law in some other
way or a Court will find you at fault.

Guidance
The Regulations and Approved Code of Practice (ACOP) are accompanied by guidance. Following the guidance is not
compulsory and you are free to take other action. But if you do follow the guidance you will normally be doing enough to
comply with the law. Health and safety inspectors seek to secure compliance with the law and may refer to this guidance as
illustrating good practice.

Presentation
The ACOP text is set out in **bold**, the accompanying guidance is in normal type, and the text of the Regulations is in *italics*.
Coloured borders also indicate each section clearly.

Contents

Introduction

1   The Ionising Radiations Regulations 2017 (IRR17) set out your minimum legal duties and the Approved Code of Practice (ACOP) text and guidance within this publication give practical advice on how to comply with those Regulations. The format of this publication is designed to clearly distinguish between the Regulations, the ACOP and the guidance.

2   This publication is intended for use by employers, but it is also relevant to employees and contractors who work with ionising radiation, radiation protection advisers and radiation protection supervisors, as well as general health and safety officers. It can also be used by self-employed people who work with ionising radiation and have certain duties under these Regulations both as an employer and as an employee.

3   The ACOP text, Regulations and guidance reflect the requirements of the Ionising Radiations Regulations 2017[1] which implement the Basic Safety Standards Directive 2013/59/Euratom[2] in Great Britain (Northern Ireland publishes separate regulations). The IRR17 came into force on 1 January 2018 and they replaced the Ionising Radiations Regulations 1999 (IRR99).

4   On 6 February 2018, the IRR17 were amended by the Ionising Radiation (Medical Exposure) Regulations 2017 (IRMER17). The main amendment was to remove regulation 33 (equipment used for medical exposure) from IRR17, as provision for medical equipment has now been included in IRMER17. This ACOP text, Regulations and guidance reflect the IRR17 as amended by IRMER17.

The Basic Safety Standards Directive

5   The 2013 Basic Safety Standards Directive (referred to as BSSD within this publication) brings five directives and an EU commission recommendation into one Directive:

■   Basic Safety Standards Directive 96/29/Euratom;
■   Medical Exposures Directive 97/43/Euratom;
■   Outside Workers Directive 90/641/Euratom;
■   Control of high-activity sealed radioactive sources and orphan sources 2003/122/Euratom;
■   Public Information Directive 89/618/Euratom;
■   Radon Commission Recommendation 90/143/Euratom.

6   The BSSD lays down requirements for protection against the dangers arising from exposure to ionising radiation. The aims of the Directive are to make sure:

■   minimum standards for protection against ionising radiation are introduced across all member states;
■   dutyholders minimise, so far as is reasonably practicable, the risks from ionising radiation to which workers, the public and others may be exposed;
■   risks from ionising radiation are controlled.

Scope of the revised Regulations

7   Regulation 3 of IRR17 sets out the scope of application of the Regulations. They apply to:

- any practice as defined in regulation 2(1) including work with materials containing naturally occurring radionuclides;
- any work (other than a practice) carried on in an atmosphere containing radon 222 gas at an annual average activity concentration in air exceeding 300 Bq m^{-3}.

8   Regulation 3 also sets out provisions that do not apply to people undergoing medical exposures. Such exposures are regulated by the Department of Health and Social Care and the Devolved Administrations.

Self-employed people

9   Regulation 2(2) extends references to the terms employer and employee, in certain circumstances, to include self-employed people. For example, self-employed people may be required to:

- carry out an assessment under regulation 8;
- provide control measures to restrict exposures under regulation 9;
- arrange for their own training under regulation 15;
- designate themselves as classified persons under regulation 21;
- make arrangements with an approved dosimetry service (ADS) for assessment and recording of their doses under regulation 22;
- obtain and keep a radiation passbook up to date if they are an outside worker;
- make sure that they use a dosemeter provided by an ADS as in regulation 35.

10   Although only the courts can give an authoritative interpretation of law, in considering the application of these Regulations, ACOP and guidance to people working under another's direction, the following should be considered:

- if people working under the control and direction of others are treated as self-employed for tax and national insurance purposes they may still be treated as their employees for health and safety purposes. It may therefore be necessary to take appropriate action to protect them;
- if any doubt exists about who is responsible for the health and safety of a worker this could be clarified and included in the terms of a contract. However, remember that a legal duty under section 3 of the Health and Safety at Work etc Act 1974 (HSWA)[3] cannot be passed on by means of a contract and there will still be duties towards others under section 3 of HSWA. If you employ such workers on the basis that they are responsible for their own health and safety, you should seek legal advice before doing so.

About this publication

11   This publication has been revised to take account of IRR17, as amended by IRMER17 (see paragraph 4 above). It sets out the Regulations, ACOP and associated guidance. It forms a framework which, if followed, makes sure exposure to ionising radiation arising from work activities is kept as low as reasonably practicable and does not exceed the dose limits specified within the Regulations.

12   The changes, which are summarised below, have been widely consulted on. They include:

- lowering the dose limit to the lens of the eye;
- flexibility for five-year averaging for dose limit to lens of the eye, subject to conditions specified by HSE;
- change to the radon reference level. The IRR99 radon reference level was over a 24-hour period, while BSSD expresses the reference level on an annual basis. Calculations show that the IRR99 reference level is broadly equivalent to the annual average reference level in IRR17;
- introducing a three-tier system of notification, registration and consent that replaces the IRR99 requirement for notification and prior authorisation;
- change to the requirement for notification, which for some radionuclides is at a lower threshold than in IRR99;
- broadening the scope of the definition of an outside worker so that it includes both classified and non-classified workers;
- change to the dose record retention period from 50 years to not less than 30 years after the last day of work;
- a requirement to put procedures in place to estimate doses to members of the public;
- a change to remove the requirement for a registered medical practitioner to be appointed 'in writing' for the purposes of these Regulations;
- introducing a requirement for authorisation of the annual whole-body dose limit in special cases – HSE or Office for Nuclear Regulation (ONR) may authorise the application of an effective dose limit of 100 mSv over five years (with no more than 50 mSv in a single year) rather than dutyholders only giving prior notification;
- recording and analysis of significant events, ie radiation accidents;
- removing the subsidiary dose limit for the abdomen of a woman of reproductive capacity;
- removing references to 'radiation employers', a term that has previously caused confusion, and replacing it where appropriate, with reference to an employer who works with ionising radiation;
- those required to accommodate other legislative, standard or guidance changes.

About ACOPs

13   Approved Codes of Practice are approved by the HSE Board with the consent of the Secretary of State (see Appendix 1: Notice of Approval for details).

14   The ACOP paragraphs describe preferred or recommended methods that can be used (or standards to be met) to comply with the Regulations and the duties imposed by the HSWA. The accompanying guidance also provides advice on achieving compliance, or it may give information of a general nature, including explanation of the requirements of the law, more specific technical information or references to further sources of information.

15   The legal status of ACOP and guidance text is given on the copyright page.

PART 1 Preliminary

Regulation 1 Citation and commencement

Regulation	1

(1)   These Regulations may be cited as the Ionising Radiations Regulations 2017.

(2)   They come into force on 1st January 2018.

Regulation 2 Interpretation

Regulation	2(1)

(1)   In these Regulations –

"the 1974 Act" means the Health and Safety at Work etc. Act 1974;

"accelerator" means an apparatus or installation in which particles are accelerated and which emits ionising radiation with an energy higher than 1 MeV;

"appointed doctor" means a registered medical practitioner who meets such recognition criteria as may from time to time be specified in writing by the Executive;

"approved" means approved for the time being in writing for the purposes of these Regulations by the Executive or the ONR (as the case may be) and published in such form as that body considers appropriate;

"approved dosimetry service" means a dosimetry service approved in accordance with regulation 36;

"authorised defence site" has the meaning given by regulation 2(1) of the Health and Safety (Enforcing Authority) Regulations 1998;

"calendar year" means a period of 12 months beginning with the 1st January;

"classified outside worker" means a classified person who carries out services in the controlled area of any employer (other than the controlled area of their own employer);

"classified person" means –

> *(a)   a person designated as such pursuant to regulation 21(1); and*
> *(b)   in the case of a classified outside worker employed by an undertaking in Northern Ireland or in another member State, a person who has been designated as a Category A exposed worker within the meaning of Article 40 of the Directive;*

"carers and comforters" means individuals knowingly and willingly incurring an exposure to ionising radiation by helping, other than as part of their occupation, in the support and comfort of individuals undergoing or having undergone medical exposure;

Regulation	2(1)

"contamination" means the unintended or undesirable presence of radioactive substances on surfaces or within solids, liquids or gases or on the human body, and "contaminated" is to be construed accordingly;

"controlled area" means –

    (a) in the case of an area situated in Great Britain, an area which has been so designated in accordance with regulation 17(1); and

    (b) in the case of an area situated in Northern Ireland or in another member State, an area subject to special rules for the purposes of protection against ionising radiation and to which access is controlled as specified in Article 37 of the Directive;

"the Directive" means Council Directive 2013/59/ Euratom laying down basic safety standards for protection against the dangers arising from exposure to ionising radiation, and repealing Directives 89/618/ Euratom, 90/641/Euratom, 96/29/Euratom, 97/43/Euratom and 2003/122/Euratom;

"dose" means, in relation to ionising radiation, any dose quantity or sum of dose quantities mentioned in Schedule 3;

"dose assessment" means the dose assessment made and recorded by an approved dosimetry service in accordance with regulation 22;

"dose constraint" means a constraint set on the prospective doses of individuals which may result from a given radiation source;

"dose limit" means, in relation to persons of a specified class, the limit on effective dose or equivalent dose specified in Schedule 3 in relation to a person of that class;

"dose rate" means, in relation to a place, the rate at which a person or part of a person would receive a dose of ionising radiation from external radiation if that person were at that place, being a dose rate at that place averaged over one minute;

"dose record" means, in relation to a person, the record of the doses received by that person as a result of that person's exposure to ionising radiation, being the record made and maintained on behalf of their employer by the approved dosimetry service in accordance with regulation 22;

"employment medical adviser" means an employment medical adviser appointed under section 56 of the 1974 Act;

"external radiation" means, in relation to a person, ionising radiation coming from outside the body of that person;

"extremities" means a person's hands, forearms, feet and ankles;

"health record" means, in relation to an employee, the record of medical surveillance of that employee maintained by the employer in accordance with regulation 25(3);

"high-activity sealed source" means a sealed source for which the quantity of the radionuclide is equal to or exceeds the relevant quantity value set out in Part 4 of Schedule 7;

"industrial irradiation" means the use of ionising radiation to sterilise, process or alter the structure of products or materials;

Regulation 2(1)

"industrial radiography" means the use of ionising radiation for non-destructive testing purposes where an image of the item under test is formed (but excluding any such testing which is carried out in a cabinet which a person cannot enter);

"internal radiation" means in relation to a person, ionising radiation coming from inside the body of that person;

"ionising radiation" means the transfer of energy in the form of particles or electromagnetic waves of a wavelength of 100 nanometres or less or a frequency of 3×10^{15} hertz or more capable of producing ions directly or indirectly;

"local rules" means rules made pursuant to regulation 18(1);

"maintained", where the reference is to maintaining plant, apparatus, equipment or facilities, means maintained in an efficient state, in efficient working order and good repair;

"medical exposure" means the exposure to ionising radiation of –

(a) patients and asymptomatic individuals as part of their own medical diagnosis or treatment;
(b) individuals as part of health screening programmes;
(c) patients or other persons voluntarily participating in medical or biomedical, diagnostic or therapeutic, research programmes;
(d) individuals undergoing non-medical imaging using medical radiological equipment;
(e) carers and comforters;

"member State" means a member State of the European Union;

"new nuclear build site" has the meaning given by Regulation 2A of the Health and Safety (Enforcing Authority) Regulations 1998;

"non-classified outside worker" means a person who is not a classified person who carries out services in the supervised or, pursuant to regulation 19(3)(c), controlled area of any employer (other than the supervised or controlled area of their own employer);

"nuclear premises" means premises which are or are on –

(a) a GB nuclear site (within the meaning given by section 68 of the Energy Act 2013);
(b) an authorised defence site;
(c) a new nuclear build site; or
(d) a nuclear warship site;

"nuclear warship site" has the meaning given by regulation 2B of the Health and Safety (Enforcing Authority) Regulations 1998;

"the ONR" means the Office for Nuclear Regulation;

"outside worker" means a classified outside worker and a non-classified outside worker;

Guidance 2(1)

16   An outside worker is any person who is carrying out services in a controlled area or supervised area but who does not have an individual contract of employment with the employer responsible for that area (as distinct from any contract for service between their own employer and the employer responsible for

the area). 'Carrying out services' implies providing a benefit to the employer responsible for the controlled or supervised area.

17   The intention of the definition is that all outside workers, including non-classified outside workers, have the same level of protection as other employees (those formally employed by the organisation) in relation to training, instruction, protective equipment, dose monitoring and entry to controlled and supervised areas.

18   Employees who are based at one site but who visit other sites which are all part of their employer's organisation are not outside workers. However, where each site is controlled by a different subsidiary company (ie a separate legal entity) then the employees will be outside workers on sites other than their base.

19   Self-employed contractors have responsibilities both as outside workers and as employers when they perform services in the controlled or supervised areas of other employers (see regulation 2(2)).

"overexposure" means any exposure of a person to ionising radiation to the extent that the dose received by that person causes a dose limit relevant to that person to be exceeded or, in relation to regulation 27(2), causes a proportion of a dose limit relevant to any employee to be exceeded;

"practice" means work involving –

(a)   the production, processing, handling, disposal, use, storage, holding or transport of radioactive substances; or

(b)   the operation of any electrical equipment emitting ionising radiation and containing components operating at a potential difference of more than 5kV, which can increase the exposure of individuals to ionising radiation;

20   Employers that work with ionising radiation in the context of these Regulations, carry out:

(a)   a practice (see definition in regulation 2(1)); or

(b)   work in places where the radon gas concentration exceeds the values in regulation 3(1)(b).

"radiation accident" means an accident where immediate action would be required to prevent or reduce the exposure to ionising radiation of employees or any other persons;

"radiation generator" means a device capable of generating ionising radiation such as x-rays, neutrons, electrons or other charged particles;

"radiation passbook" means –

(a)   in the case of a classified outside worker employed by an employer in Great Britain –
(i)   a passbook approved by the Executive for the purpose of these Regulations; or
(ii)   a passbook to which paragraph 9 of Schedule 8 (transitional provisions) applies; and

(b)   in the case of a classified outside worker employed by an employer in Northern Ireland or in another member State, a passbook authorised by the competent authority for Northern Ireland or that member State, as the case may be;

| Regulation | 2(1) |

"radiation protection adviser" means an individual who, or a body which, meets such criteria of competence as may from time to time be specified in writing by the Executive;

"radioactive material" means material incorporating radioactive substances;

"radioactive source" means an entity incorporating a radioactive substance (or substances) for the purpose of utilising the radioactivity of that substance (or substances);

"radioactive substance" means any substance which contains one or more radionuclides whose activity cannot be disregarded for the purposes of radiation protection;

| ACOP | 2(1) |

21 For a substance used in a practice, its activity should never be disregarded for the purposes of radiation protection where:

(a) **that activity concentration exceeds the values set out in column 2 of Part 1 of Schedule 7, subject to the quantity of the substance also exceeding the values set out in column 3 of Part 1 of Schedule 7 (for artificial radionuclides and naturally occurring radionuclides which are processed for their radioactive, fissile or fertile properties); or**

(b) **that activity concentration exceeds the values set out in column 2 of Part 2 of Schedule 7, subject to the quantity of the substance also exceeding the values set out in column 3 of Part 2 of Schedule 7 (for naturally occurring radionuclides which are not processed for their radioactive, fissile or fertile properties).**

| Guidance | 2(1) |

22 Whether a particular material is considered to be a radioactive substance within the meaning of the Regulations depends on a judgement about its radiological impact. For substances used in practices, this Approved Code of Practice gives practical advice on situations where substances should always be regarded as radioactive substances. Below these activity levels, substances may still be considered 'radioactive' for the purposes of the Regulations, depending on the radionuclide concerned, the way in which they are used and the particular circumstances of exposure. For example, a substance used only in solid form may have an activity that can be disregarded, but the same substance, used in such a way that people can inhale that material as a fine dust, might need to be treated as a radioactive substance for the purposes of these Regulations.

Work activities with naturally occurring radioactive materials

23 Judging what is considered to be a radioactive substance within the meaning of the Regulations can be particularly difficult for materials containing naturally occurring radionuclides. Processing of such materials can concentrate certain radionuclides and lead to the potential for significant exposure at particular stages of a process. The two main exposure routes are from direct exposure to external radiation from bulk quantities (often held in store) and from inhalation during work in dusty operations.

24 Some processes which may involve exposures from naturally occurring radioactive materials are:

(a) oil and gas extraction, where scale in pipes and vessels may contain significant amounts of uranium and thorium and their decay products including radium;

(b) some forms of metal processing, where refractory materials or feed ores may contain naturally occurring radioactive materials and where

Guidance   **2(1)**

radionuclides can volatilise and condense, or concentrate in the product or the slags to enhance activity levels;

(c) thorium alloy manufacture, for example for aircraft parts and the use of thoriated products, such as special types of welding electrodes;

(d) coal-fired power plants in maintaining boilers;

(e) production of phosphate fertilisers;

(f) cement production, maintenance of clinker ovens;

(g) tin/lead/copper smelting;

(h) ground water filtration facilities;

(i) geothermal energy production;

(j) primary iron production;

(k) titanium dioxide pigment production;

(l) thermal phosphorus production;

(m) zircon and zirconium industry;

(n) phosphoric acid production;

(o) mining of ores other than uranium;

(p) extraction of rare earths from monazite;

(q) processing of niobium/tantalum ore;

(r) extraction of china clay.

Regulation   **2(1)**

"relevant doctor" means an appointed doctor or an employment medical adviser;

"sealed source" means a radioactive source whose structure is such as to prevent, under normal conditions of use, any dispersion of radioactive substances into the environment, but it does not include any radioactive substance inside a nuclear reactor or any nuclear fuel element;

"supervised area" means an area which has been so designated by the employer in accordance with regulation 17(3);

"trainee" means a person aged 16 years or over (including a student) who is undergoing instruction or training which involves operations which would, in the case of an employee, be work with ionising radiation;

"transport" means, in relation to a radioactive substance, carriage of that substance on a road within the meaning of, in relation to England and Wales, section 192 of the Road Traffic Act 1988 and, in relation to Scotland, section 151 of the Roads (Scotland) Act 1984 or through another public place (whether on a conveyance or not), or by rail, inland waterway, sea or air and, in the case of transport on a conveyance, a substance is deemed as being transported from the time that it is loaded onto the conveyance for the purpose of transporting it until it is unloaded from that conveyance, but a substance is not to be considered as being transported if –

(a) it is transported by means of a pipeline or similar means; or

(b) it forms an integral part of a conveyance and is used in connection with the operation of that conveyance;

"work with ionising radiation" means work to which these Regulations apply by virtue of regulation 3(1).

Guidance   **2(1)**

Work with ionising radiation

25 Work with ionising radiation means:

(a) the production, processing, handling, disposal use, storage, holding or transport of radioactive substances;

(b) the operation of electrical equipment emitting ionising radiation and containing components operating at energies above 5 kV (see definition of practice in regulation 2(1));

(c) work in places where the radon gas concentration exceeds the values in regulation 3(1)(b).

Regulation 2(2)

(2) In these Regulations, any reference to –

(a) an employer includes a reference to a self-employed person and any duty imposed by these Regulations on an employer in respect of that employer's employee extends to a self-employed person in respect of themselves;

(b) an employee includes a reference to –
  (i) a self-employed person, and
  (ii) a trainee who but for the operation of this sub-paragraph and paragraph (3) would not be classed as an employee;

(c) exposure to ionising radiation is a reference to exposure to ionising radiation arising from work with ionising radiation;

(d) a person entering, remaining in or working in a controlled or supervised area includes a reference to any part of a person entering, remaining in or working in any such area.

Guidance 2(2)

Duties of self-employed people

26 A self-employed person who works with ionising radiation will have certain duties under these Regulations both as an employer and as an employee (see guidance to regulation 35 and the Introduction).

Regulation 2(3)–(5)

(3) For the purposes of these Regulations and Part I of the 1974 Act –

(a) the word "work" is extended to include any instruction or training which a person undergoes as a trainee and the meaning of "at work" is extended accordingly; and

(b) a trainee, while undergoing instruction or training in respect of work with ionising radiation, is to be treated as the employee of the person whose undertaking (whether for profit or not) is providing that instruction or training and that person is to be treated as the employer of that trainee except that the duties to the trainee imposed upon the person providing instruction or training will only extend to matters under the control of that person.

(4) In these Regulations, where reference is made to a quantity or concentration specified in Schedule 7, that quantity or concentration is to be treated as being exceeded if –

(a) where only one radionuclide is involved –
  (i) the quantity of that radionuclide exceeds the quantity specified in the appropriate entry in Parts 1, 2 or 4 of Schedule 7; or
  (ii) the concentration of that radionuclide exceeds the concentration specified in the appropriate entry in Parts 1 or 2 of Schedule 7; or

(b) where more than one radionuclide is involved, the quantity or concentration ratio calculated in accordance with Part 3 of Schedule 7 exceeds one.

(5) Nothing in these Regulations is to be construed as preventing a person from entering or remaining in a controlled area or a supervised area where that person enters or remains in any such area –

Regulation 2(5)–(6)

(a) in the due exercise of a power of entry conferred on that person by or under any enactment; or

(b) for the purpose of undergoing a medical exposure.

(6) In these Regulations –

(a) any reference to an effective dose means the sum of the effective dose to the whole body from external radiation and the committed effective dose from internal radiation; and

(b) any reference to equivalent dose to a human tissue or organ includes the committed equivalent dose to that tissue or organ from internal radiation.

Regulation 3 Application

Regulation   3(1)

(1) Subject to the provisions of this regulation and to regulation 5(1), these Regulations apply to –

(a) any practice; and

(b) any work (other than a practice) carried on in an atmosphere containing radon 222 gas at an annual average activity concentration in air exceeding 300 Bq m^{-3}.

Guidance   3(1)

27   There are two basic types of work to which the Regulations apply. A practice, the most common type, is defined in regulation 2(1) and covers the normal activities associated with use of radiation sources, such as power generation and industrial radiography. A practice also includes work with materials containing naturally occurring radionuclides. Regulation 3(1)(b) covers work in underground facilities such as mines, caves and tunnels and other workplaces where the construction of the workplace and the ventilation provided is insufficient to keep the concentration below the specified levels.

Regulation 3(2)–(5)

(2) The following regulations do not apply where the only work being undertaken is that referred to in paragraph (1)(b), namely regulations 24, 28 to 31, 33 and 34.

(3) The following regulations do not apply in relation to persons undergoing medical exposures, namely regulations 8, 9, 12, 17 to 19, 24, 26, 32(1) and 35(1).

(5) In the case of a classified outside worker (working in a controlled area situated in Great Britain) employed by an employer established in Northern Ireland or in another member State, it is sufficient compliance with regulation 22 (dose assessment and recording) and regulation 25 (medical surveillance) if the employer complies with –

(a) where the employer is established in Northern Ireland, regulations 21 and 24 of the Ionising Radiations Regulations (Northern Ireland) 2000 or any other provision made for the purpose of implementing the relevant parts of Chapter VI of the Directive in Northern Ireland; or

(b) where the employer is established in another member State, the legislation in that State implementing the relevant parts of Chapter VI of the Directive where such legislation exists.

Note - On 6 February 2018, regulation 3(4) was deleted from IRR17. Provisions relating to medical equipment are now contained in the Ionising Radiation (Medical Exposure) Regulations 2017.

28   This requirement avoids the need for a further pre-classification medical examination when an outside worker arrives in Great Britain from a member state (or Northern Ireland). It also means that an outside worker from a member state does not have to wear an additional dosemeter, issued by an approved dosimetry service (ADS) in Great Britain, for the purpose of dose assessment.

Regulation 4 Duties under the Regulations

Regulation   4(1)–(4)

(1)   Any duty imposed by these Regulations on an employer in respect of the exposure to ionising radiation of persons other than that employer's employees is imposed only in so far as the exposure of those persons to ionising radiation arises from work with ionising radiation undertaken by that employer.

(2)   Duties under these Regulations imposed upon the employer are also imposed upon any person who is –

(a)   a mine operator; or
(b)   the operator of a quarry,

in so far as those duties relate to the mine or part of the mine of which that person is the mine operator or the quarry of which that person is the operator and to matters within that person's control.

(3)   Subject to regulation 5(1)(c), 6(2)(c) and (d) and 7(1)(h), duties under these Regulations imposed upon the employer are imposed on the holder of a nuclear site licence under the Nuclear Installations Act 1965 in so far as those duties relate to the licensed site.

(4)   In this regulation –

(a)   "mine operator" has the meaning given by regulation 2(1) of the Mines Regulations 2014;
(b)   "operator", in relation to the operator of a quarry, has the meaning given by regulation 2(1) of the Quarries Regulations 1999.

PART 2 General principles and procedures

Regulation 5 Notification of certain work

Regulation	5(1)

(1)   This regulation applies to work with ionising radiation except –

(a)   work arising from the carrying out of a registrable practice under regulation 6 or a specified practice requiring consent under regulation 7;

(b)   work specified in Schedule 1; and

(c)   work carried on at a site licensed under section 1 of the Nuclear Installations Act 1965.

Guidance	5(1)

29   Regulatory control of work with ionising radiation is proportionate to the size and likelihood of exposures resulting from that work. When considered together, regulations 5, 6 and 7 form a three-tier, risk-based approach to how work with ionising radiation is categorised. Regulation 5 relates to notifications, the lowest tier of the approach.

30   Notification applies to any work with ionising radiation that does not require registration (regulation 6) or consent (regulation 7), as long as the work is not listed in Schedule 1.

31   In practice, this means that the following work with ionising radiation requires notification:

(a)   work with 1000 kg or less of radioactive material containing artificial radionuclides or radioactive material containing naturally occurring radionuclides which are processed for their radioactive, fissile or fertile properties:

(i)   if the concentration of the radioactive substance is above the value in column 2, but does not exceed that in column 4 of Schedule 7, Part 1, and the quantity of radioactivity exceeds the value in column 3 of Schedule 7, Part 1;

(b)   work with 1000 kg or less of radioactive material containing naturally occurring radionuclides which are not processed for their radioactive, fissile or fertile properties:

(i)   if the concentration is above the value in column 2, but does not exceed that in column 4 of Schedule 7, Part 2, and the quantity of radioactivity exceeds the value in column 3 of Schedule 7, Part 2;

(c)   work carried out in an atmosphere containing radon 222 gas at an annual average concentration in air exceeding 300 Bq m^{-3}. See regulation 3(1)(b).

32   Note that work with radiation generators, such as X-ray devices, does not require notification – they require registration, at the least.

33   Where specific values for radionuclides are not detailed in Schedule 7 Part 1, values for 'other radionuclides not listed above' should be used. HSE may approve specific values for unlisted radionuclides. Part 3 of Schedule 7 explains how to calculate values for exemption from notification where more than one radionuclide is involved.

Guidance   5(1)

34   Work with ionising radiation on licensed nuclear sites does not need to be notified under this regulation.

35   See the relevant exemption flowchart for artificial radionuclides and naturally occurring radioactive material (Figures 1 and 2) for an explanation of the three-tier approach to regulation.

Figure 1 Exemption flowchart – Schedule 7 Part 1. Artificial and naturally occurring radionuclides (processed for their radioactive, fissile or fertile properties)

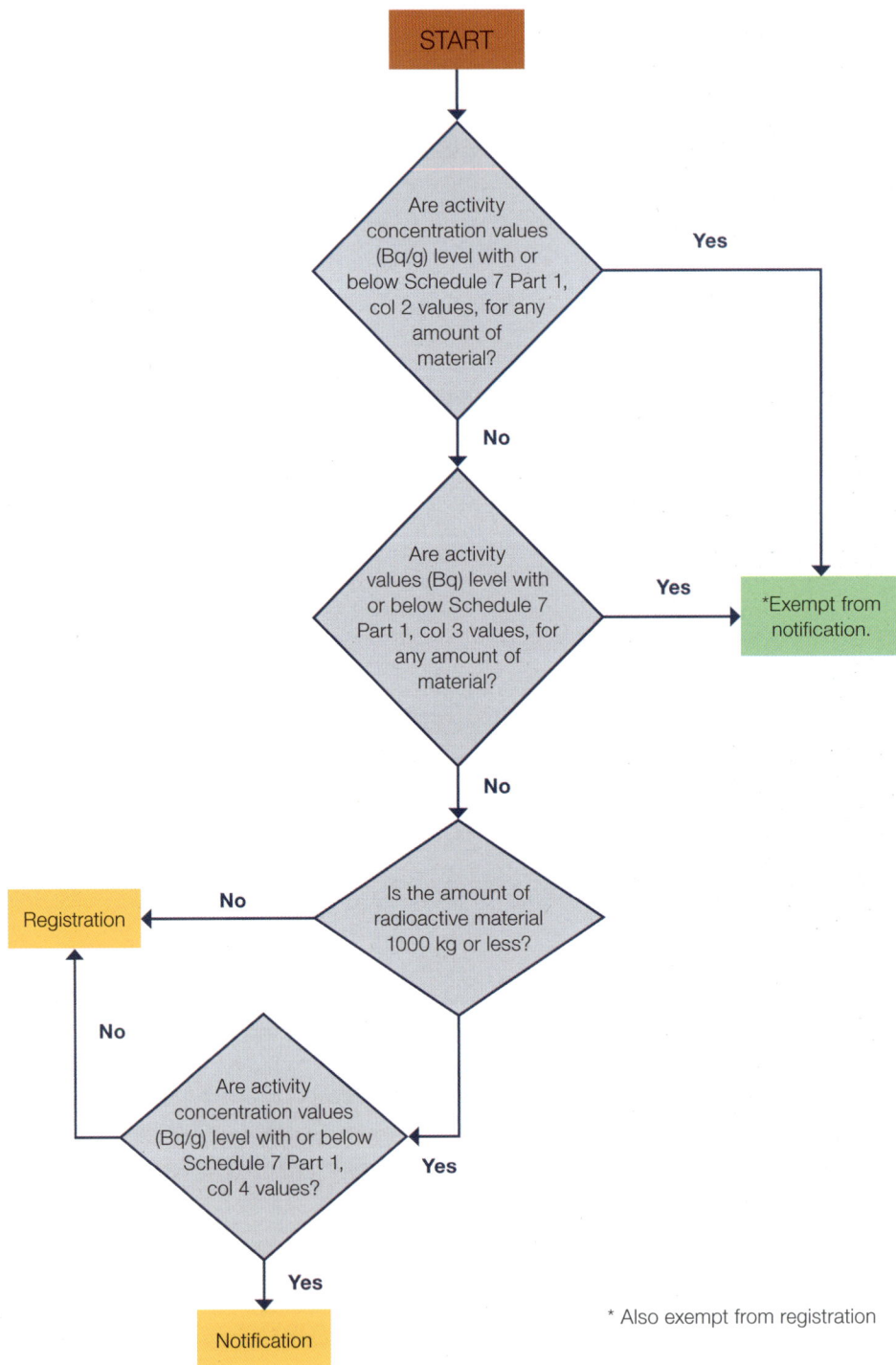

START

Are activity concentration values (Bq/g) level with or below Schedule 7 Part 1, col 2 values, for any amount of material? — Yes → *Exempt from notification.

No ↓

Are activity values (Bq) level with or below Schedule 7 Part 1, col 3 values, for any amount of material? — Yes → *Exempt from notification.

No ↓

Is the amount of radioactive material 1000 kg or less? — No → Registration

Yes ↓

Are activity concentration values (Bq/g) level with or below Schedule 7 Part 1, col 4 values? — No → Registration

Yes ↓

Notification

* Also exempt from registration

Figure 2 Exemption flowchart – Schedule 7 Part 2. Naturally occurring radionuclides (not processed for their radioactive, fissile or fertile properties)

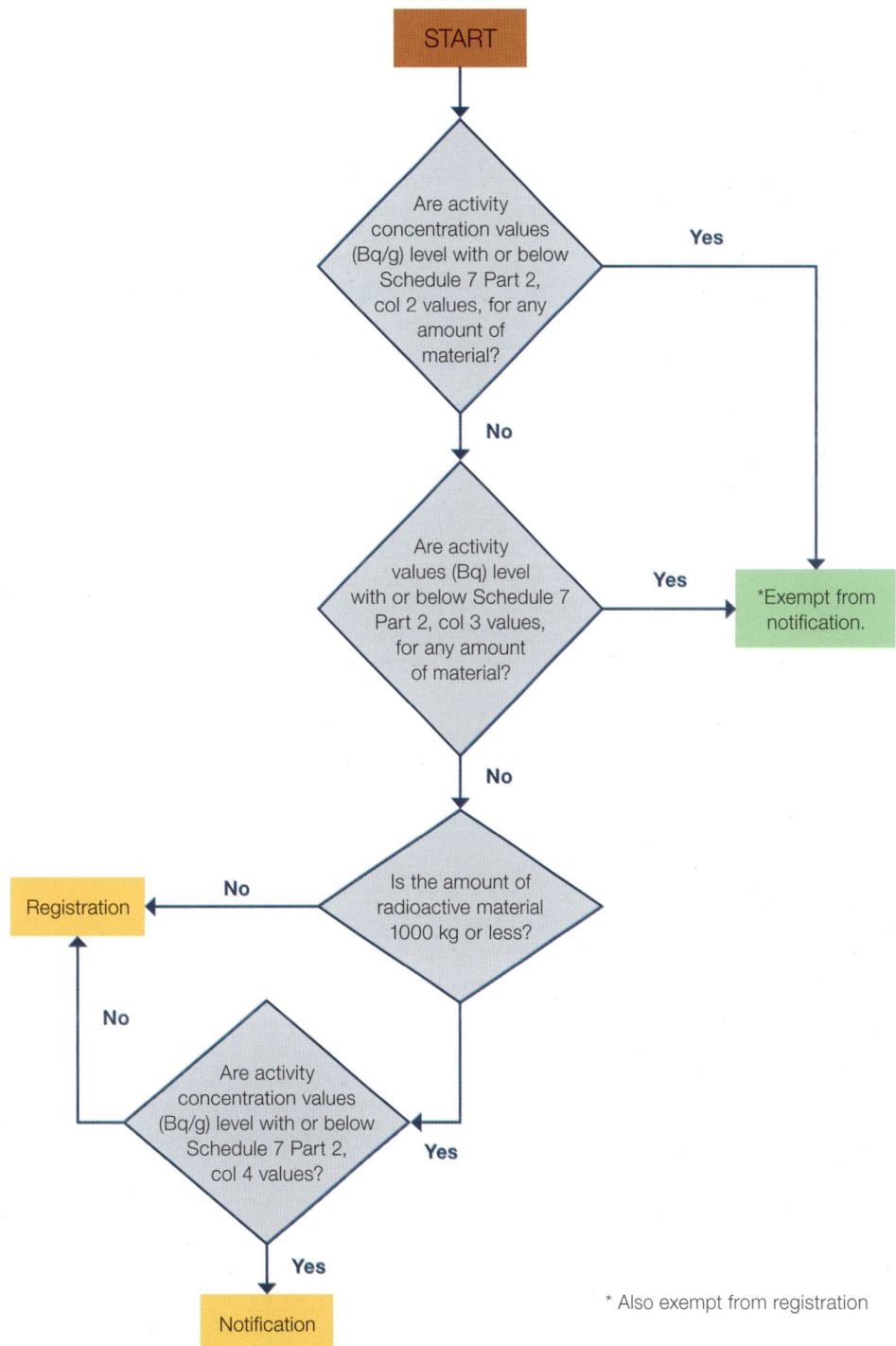

* Also exempt from registration

Regulation 5(2)

(2) Subject to paragraph 3 of Schedule 8 (which relates to transitional provisions), an employer must not carry out work with ionising radiation to which this regulation applies unless before the first occasion of commencing such work since the coming into force of this regulation the employer has notified that work to the appropriate authority in accordance with the notification procedure approved by the appropriate authority from time to time.

Guidance 5(2)

36   Employers intending to carry out work with ionising radiation that requires notification must notify HSE of their intention to work before carrying out that work via HSE's online system: www.hse.gov.uk/radiation/ionising/notification-process.htm.

37   Work requiring notification under IRR17 that has been notified to HSE under previous Ionising Radiations Regulations must be subject to a new notification submission via the online system, in accordance with the transitional arrangements detailed in Schedule 8.

Regulation 5(3)–(5)

(3) Where an employer has notified work in accordance with paragraph (2), the appropriate authority may, by notice in writing, require that employer to provide such additional particulars of that work as the appropriate authority may reasonably require in connection with the notification, and in such a case the employer must provide those particulars by such time as is specified in the notice or by such other time as the appropriate authority may subsequently agree.

(4) A notice under paragraph (3) may require the employer to notify the appropriate authority of any of those additional particulars before each occasion on which the employer commences work with ionising radiation.

(5) Where an employer has notified work in accordance with this regulation and subsequently ceases that work, or makes a material change in the work which would affect the particulars provided to the appropriate authority in connection with the notification, the employer must immediately notify the appropriate authority of that cessation or material change.

Guidance 5(5)

38   Employers must inform HSE if:

(a)   there is a significant change to the details previously notified, for example the employer changes address, or the work is being carried out at premises not under the employer's control; or

(b)   the notified work has been stopped and will not be performed again.

Regulation 5(6)

(6) In this regulation "appropriate authority" means –

(a)   *in relation to work carried on exclusively or primarily on premises which are or are on –*
(i)   *an authorised defence site;*
(ii)   *a new nuclear build site;*
(iii) *a nuclear warship site, the ONR;*
(b)   *otherwise, the Executive.*

Regulation 6 Registration of certain practices

(1) For the purposes of this regulation, all practices are registrable practices except those practices listed in paragraph (2).

(2) The following practices are not registrable practices –

(a) *a practice solely involving work with ionising radiation to which Schedule 1 applies;*

(b) *a specified practice (within the meaning of regulation 7(1));*

(c) *the operation or decommissioning of any nuclear installation;*

(d) *the operation, decommissioning or closure of any facility for the long-term storage or disposal of radioactive waste (including facilities managing radioactive waste for this purpose) where such facility is situated on a site licensed under section 1 of the Nuclear Installations Act 1965;*

(e) *any practice involving radioactive material where the amount of the radioactive material does not exceed 1,000kg and the activity concentration value of the radioactive substance in that material does not exceed the value specified in Column 4 of Part 1 of Schedule 7 (for artificial radionuclides and naturally occurring radionuclides which are processed for their radioactive, fissile or fertile properties) or Column 4 of Part 2 of Schedule 7 (for naturally occurring radionuclides which are not processed for their radioactive, fissile or fertile properties);*

(f) *any practice involving radioactive material where the amount of the radioactive material exceeds 1,000kg and the activity concentration value of the radioactive substance in that material does not exceed the value in Column 2 of Part 1 of Schedule 7 (for artificial radionuclides and naturally occurring radionuclides which are processed for their radioactive, fissile or fertile properties) or Column 2 of Part 2 of Schedule 7 (for naturally occurring radionuclides which are not processed for their radioactive, fissile or fertile properties);*

(3) Subject to paragraph 5 of Schedule 8 (which relates to transitional provisions), an employer must not carry out a registrable practice unless that employer has applied for, and has been issued with, a registration in connection with the practice by the appropriate authority.

(4) An employer applying for a registration under paragraph (3) must provide –

(a) *such information regarding the practice as is required by the registration procedure approved by the appropriate authority from time to time; and*

(b) *upon notice in writing by the appropriate authority, such other information relating to the practice as the appropriate authority may reasonably require in connection with the registration.*

(5) A registration under paragraph (3) may be issued subject to conditions (which may include a limit of time) and may be revoked in writing at any time.

39 Paragraph 30 sets out HSE's approach to regulatory control. Regulation 6 relates to registrations, the middle-risk tier of the approach to regulatory control.

Circumstances under which registration is not required

40 Registration applies to any work with ionising radiation that does not require notification (regulation 5) or consent (regulation 7), as long as the work is not listed in Schedule 1.

41 In practice, this means the following work with ionising radiation requires registration:

(a) work with a radiation generator:
(i) unless this work with a radiation generator is a specified practice requiring consent under regulation 7, or Schedule 1 applies. An X-ray device is a radiation generator;

(b) work with 1000 kg or less of radioactive material containing artificial radionuclides or radioactive material containing naturally occurring radionuclides which are processed for their radioactive, fissile or fertile properties:
(i) if the concentration is above the value in column 4 of Schedule 7, Part 1;

(c) work with 1000 kg or less of radioactive material containing naturally occurring radionuclides which are not processed for their radioactive, fissile or fertile properties:
(i) if the concentration is above the value in column 4 of Schedule 7, Part 2;

(d) work with over 1000 kg of radioactive material containing artificial radionuclides or radioactive material containing naturally occurring radionuclides which are processed for their radioactive, fissile or fertile properties:
(i) if the concentration is above the value in column 2 of Schedule 7, Part 1;

(e) work with over 1000 kg of radioactive material containing naturally occurring radionuclides which are not processed for their radioactive, fissile or fertile properties:
(i) if the concentration is above the value in column 2 of Schedule 7, Part 2.

Registerable practices

42 Employers intending to carry out work with ionising radiation that requires registration must apply to HSE via the online system before carrying out that work: www.hse.gov.uk/radiation/ionising/notification-process.htm.

43 Work requiring registration that has been notified to HSE under previous Ionising Radiations Regulations must be subject to a registration submission via the online system in accordance with the transitional arrangements identified in Schedule 8.

44 Work requiring registration with the Office for Nuclear Regulation (ONR) will need to be directed to them. Further details can be found at www.onr.org.uk.

45 Specific details of the registration procedure and application process, including the conditions for registrations, and appeals and revocation procedures, are available on HSE's website: www.hse.gov.uk/radiation/ionising/notification-process.htm.

46 Employers that do not meet the application criteria in the registration procedure will not be granted a registration and must not carry out the specified work.

(6) Where an employer has registered a practice in accordance with this regulation and subsequently ceases to carry out that practice, or makes a material change to the practice which would affect the particulars provided to the appropriate authority in connection with the registration, the employer must immediately notify the appropriate authority of that cessation or material change.

Guidance	6(6)

Reporting a material change

47 Employers must tell HSE if there has been a significant change to the details previously submitted (see regulation 5(5)), or if the work with ionising radiation requiring registration has stopped and will not be performed again.

48 Material change has the same meaning as in regulation 5(5).

Regulation	6(7)

(7) An employer who is aggrieved by –

(a) a decision of the appropriate authority refusing to issue a registration under paragraph (3) or revoking a registration under paragraph (5); or

(b) the terms of any conditions attached to a registration under paragraph (5),

may appeal to the Secretary of State.

Guidance	6(7)

49 Specific details of the registration procedure and application process, including the conditions for registrations and appeals and revocation procedures, are available on HSE's website: www.hse.gov.uk/radiation/ionising/notification-process.htm.

Regulation 6(8)–(10)

(8) Sub-sections (2) to (6) of section 44 of the 1974 Act apply for the purposes of paragraph (7) as they apply to an appeal under section 44(1) of that Act.

(9) The Health and Safety Licensing Appeals (Hearings Procedure) Rules 1974, as respects England and Wales, and the Health and Safety Licensing Appeals (Hearings Procedure) (Scotland) Rules 1974, as respects Scotland, apply to an appeal under paragraph (7) as they apply to an appeal under sub-section (1) of section 44 of the 1974 Act, but with the modification that references to a licensing authority are to be read as references to the appropriate authority.

(10) In this regulation "appropriate authority" means –

(a) in relation to practices carried out exclusively or primarily on nuclear premises, the ONR;

(b) otherwise, the Executive;

"nuclear installation" has the meaning given by regulation 26(1) of the Nuclear Installations Act 1965.

Regulation 7 Consent to carry out specified practices

Regulation	7(1)

(1) In this regulation a "specified practice" means any of the following practices –

(a) the deliberate administration of radioactive substances to persons and, in so far as the radiation protection of persons is concerned, animals for the purpose of medical or veterinary diagnosis, treatment or research;

(b) the exploitation and closure of uranium mines;

(c) the deliberate addition of radioactive substances in the production or manufacture of consumer products or other products, including medicinal products;

(d) the operation of an accelerator (except when operated as part of a practice within sub-paragraph (e) or (f) below and except an electron microscope);

(e) industrial radiography;

(f) industrial irradiation;

<table>
<tr><td>**Regulation 7(1)**</td></tr>
</table>

(g)   any practice involving a high-activity sealed source (other than one within sub-paragraph (e) or (f) above);

(h)   the operation, decommissioning or closure of any facility for the long-term storage or disposal of radioactive waste (including facilities managing radioactive waste for this purpose) but not any such facility situated on a site licensed under section 1 of the Nuclear Installations Act 1965;

(i)   practices discharging significant amounts of radioactive material with airborne or liquid effluent into the environment.

Guidance 7(1)

50   HSE's approach to regulatory control is explained in paragraph 29. Regulation 7 relates to consents, the highest-risk tier of the approach to regulation.

51   Employers that carry out work within the highest-risk tier set out within the Regulations must obtain consent to carry out such work. Only the practices listed in regulation 7(1) require consent.

52   Specific details on consents and guidance as to who these practices apply to, and how to apply for a consent, are available on HSE's website: www.hse.gov.uk/radiation/ionising/notification-process.htm.

Regulation 7(2)–(4)

(2)   Subject to paragraph 6 of Schedule 8 (which relates to transitional provisions), an employer must not carry out a specified practice unless that employer has applied for, and has been granted, a consent to carry out the practice by the appropriate authority.

(3)   An employer applying for a consent under paragraph (2) must provide –

(a)   such of the information set out in Schedule 2 as the appropriate authority may specify from time to time as necessary to determine an application for consent; and

(b)   upon notice in writing, by the appropriate authority, such other information relating to the practice as the appropriate authority may reasonably require in connection with the application for consent.

(4)   A consent under paragraph (2) may be granted subject to conditions (which may include a limit of time) and may be revoked in writing at any time.

Guidance 7(2)–(4)

53   Employers intending to carry out work with ionising radiation that requires HSE consent must apply via HSE's online system before carrying out that work www.hse.gov.uk/radiation/ionising/notification-process.htm.

54   Work requiring consent that has been notified to HSE under previous Ionising Radiations Regulations must be subject to a consent submission via the online system, in accordance with the transitional arrangements identified in Schedule 8.

55   Work requiring consent from ONR will need to be directed to them. Further details can be found at www.onr.org.uk.

Applying for consent of a specified practice

56   Specific details of the consent procedure and application process, including the conditions for registrations, consents and revocation procedures, are available on HSE's website: www.hse.gov.uk/radiation/ionising/notification-process.htm.

57   Employers that do not meet the application criteria in the consent procedure will not be granted consent to carry out the practice and must not carry out the specified work.

Regulation 7(5)–(6)	*(5) Where an employer has been granted consent under this regulation to carry out a practice and subsequently ceases to carry out that practice, or makes a material change to the practice which would affect the particulars provided to the appropriate authority in connection with the application for consent, the employer must immediately notify the appropriate authority of that cessation or material change.*

(6)   An employer who is aggrieved by –

    (a)   a decision of the appropriate authority refusing to grant a consent under paragraph (2) or revoking a consent under paragraph (4); or

    (b)   the terms of any conditions attached to a consent under paragraph (4),

may appeal to the Secretary of State.

Guidance 7(5)

Reporting a material change

58   Employers must tell HSE if there is a significant change to the details previously submitted (see regulation 5(5)), or if the work with ionising radiation requiring consent has stopped and will not be performed again.

59   Material change has the same meaning as in regulation 5(5).

60   Material change could also mean the decay of high-activity sealed sources below the levels referenced in Schedule 7, Part 4. In this instance, a registration would be required.

Regulation 7(7)–(9)

(7)   Sub-sections (2) to (6) of section 44 of the 1974 Act apply for the purposes of paragraph (6) as they apply to an appeal under section 44(1) of that Act.

(8)   The Health and Safety Licensing Appeals (Hearings Procedure) Rules 1974, as respects England and Wales, and the Health and Safety Licensing Appeals (Hearings Procedure) (Scotland) Rules 1974, as respects Scotland, apply to an appeal under paragraph (6) as they apply to an appeal under sub-section (1) of section 44 of the 1974 Act, but with the modification that references to a licensing authority are to be read as references to the appropriate authority.

(9)   In this regulation "appropriate authority" has the meaning given in regulation 6(10).

Guidance 7(9)

Appeals

61   Specific details of the consent procedure and application process, including the conditions for consents and appeals and revocation procedures, are available on HSE's website:
www.hse.gov.uk/radiation/ionising/notification-process.htm.

Regulation 8 Radiation risk assessments

Regulation 8(1)–(2)

(1)   An employer, before commencing a new activity involving work with ionising radiation in respect of which no risk assessment has been made by that employer, must make a suitable and sufficient assessment of the risk to any employee and other person for the purpose of identifying the measures the employer needs to take to restrict the exposure of that employee or other person to ionising radiation.

(2)   Without prejudice to paragraph (1), an employer must not carry out work with ionising radiation unless it has made an assessment sufficient to demonstrate that –

Regulation 8(2)–(4)

(a)   all hazards with the potential to cause a radiation accident have been identified; and

(b)   the nature and magnitude of the risks to employees and other persons arising from those hazards have been evaluated.

(3)   Where the assessment made for the purposes of this regulation shows that a radiation risk to employees or other persons exists from an identifiable radiation accident, the employer who is subject to the obligation in paragraph (1) to make the risk assessment must take all reasonably practicable steps to –

(a)   prevent any such accident;

(b)   limit the consequences of any accident which does occur; and

(c)   provide employees with the information, instruction, training and equipment necessary to restrict their exposure to ionising radiation.

(4)   The requirements of this regulation are without prejudice to the requirements of regulation 3 (Risk assessment) of the Management of Health and Safety at Work Regulations 1999.

Guidance 8(1)–(4)

General advice on radiation risk assessment

62   Previously known as a prior risk assessment, the radiation risk assessment complements the requirements of regulation 3 of the Management of Health and Safety at Work Regulations 1999 ('the Management Regulations'). You can find more guidance on HSE's website at www.hse.gov.uk/risk/controlling-risks.htm.

63   A new activity involving work with ionising radiation must not begin until a radiation risk assessment has been completed. The risk assessment must be:

(a)   recorded (if there are five or more employees);

(b)   kept up to date;

(c)   discussed with workers and others who could be affected by the risks identified.

64   Regulation 5 of the Management Regulations also requires arrangements to be made for the effective planning, organisation, control, monitoring and review of preventive and protective measures.

65   If employers are satisfied that the intended work is already covered by a suitable and sufficient risk assessment which was carried out for the purposes of the Management Regulations, nothing further needs to be done to satisfy regulation 8(1) of IRR17. Employers do not need to carry out a further risk assessment on each occasion before the activity starts if the working conditions are the same.

66   A suitable and sufficient radiation risk assessment made under IRR17 for any new activities will satisfy the requirements of regulation 3 of the Management Regulations, as far as radiation protection is concerned. Be careful not to introduce other radiological and conventional risks associated with alternative techniques to satisfy both IRR17 and regulation 3 of the Management Regulations. For example, some control methods for restricting exposure to ionising radiation by use of distance and shielding might pose unacceptable risks of falls or back strain.

67   Employers carrying out other work not involving ionising radiation at the premises of an employer working with ionising radiation (eg general maintenance or cleaning) must co-operate with each other. They must take account of matters identified in the other employer's risk assessment when making or reviewing their own risk assessment to satisfy the Management Regulations.

68   Regulation 8(1) does not apply to the protection of those undergoing a medical examination, diagnosis or treatment. However, it does apply to:

(a)   the protection of staff who carry out those exposures;
(b)   members of the public;
(c)   other people.

Responsibility for undertaking the radiation risk assessment

69   The employer who carries out the work with ionising radiation is responsible for making sure that the radiation risk assessment is completed.

Nature of radiation risk assessment for new activities

70   Where an employer is required to carry out a radiation risk assessment, the following matters need to be considered, where they are relevant:

(a)   the nature of the sources of ionising radiation to be used, or likely to be present, including accumulation of radon in the working environment;
(b)   estimated radiation dose rates to which anyone can be exposed;
(c)   the likelihood of contamination arising and being spread;
(d)   the results of any previous personal dosimetry or area monitoring relevant to the proposed work;
(e)   advice from the manufacturer or supplier of equipment about its safe use and maintenance;
(f)   engineering control measures and design features already in place, or planned;
(g)   any planned systems of work;
(h)   estimated levels of airborne and surface contamination likely to be encountered;
(i)   the effectiveness and the suitability of PPE to be provided;
(j)   the extent of unrestricted access to working areas where dose rates or contamination levels are likely to be significant;
(k)   possible accident situations, their likelihood and potential severity;
(l)   the consequences of possible failures of control measures – such as electrical interlocks, ventilation systems and warning devices – or systems of work;
(m)   steps to prevent identified accidents, or limit their consequences.

71   This radiation risk assessment will help the employer to decide:

(a)   the action needed to make sure the radiation exposure of all people is kept as low as reasonably practicable (regulation 9(1));
(b)   the steps necessary to achieve this control of exposure by the use of engineering controls, design features, safety devices and warning devices (regulation 9(2)(a)) and, in addition, to develop systems of work (regulation 9(2)(b));
(c)   whether it is appropriate to provide PPE and if so, what type is adequate and suitable (regulation 9(2)(c));
(d)   whether it is appropriate to establish any dose constraints for planning or design purposes and if so, what values will be used (regulation 9(4));
(e)   the need to alter the working conditions of any employee who declares they are pregnant or breastfeeding (regulation 9(6));
(f)   an appropriate investigation level to check that exposures are being restricted as far as reasonably practicable (regulation 9(8));

ACOP   **8**

(g) the maintenance and testing schedules required for the control measures selected (regulation 11);

(h) what contingency plans are necessary to address reasonably foreseeable accidents (regulation 13);

(i) the training needs of classified and non-classified employees (regulation 15);

(j) the need to designate specific areas as controlled or supervised areas and to specify local rules (regulations 17 and 18);

(k) the actions needed to make sure access is restricted and other specific measures are put in place in controlled or supervised areas (regulation 19);

(l) the need to designate certain employees as classified persons (regulation 21);

(m) the content of a suitable programme of dose assessment for employees designated as classified persons and for others who enter controlled areas (regulations 19 and 22);

(n) the requirements for the leak testing of radioactive sources (regulation 28);

(o) the responsibilities of managers and workers (including outside workers) for ensuring compliance with these regulations;

(p) an appropriate programme of monitoring or auditing of arrangements to check the requirements of these regulations are being met.

Guidance   **8(1)–(4)**

72 When conducting a radiation risk assessment, employers must consult a radiation protection adviser (RPA) about the matters to be considered (see paragraph 249). Appointed safety representative(s) must also be consulted and, where appropriate, any established safety committee about the introduction of new measures at the workplace which may affect health and safety (regulation 4(a) of the Safety Representatives and Safety Committees Regulations 1977).[4] Where there is no appointed safety representative, the employer must consult employees, as required by the Health and Safety (Consultation with Employees) Regulations 1996.[5]

73 Employers must take account of the risks arising from radiation exposure to those who are pregnant or breastfeeding and, in particular, the likely doses to the foetus or the breastfed infant. The assessment should also, where appropriate, take into account any particular risks to young people resulting from their inexperience, lack of awareness of risks, and possible immaturity (regulation 3(4) of the Management Regulations).

74 The radiation risk assessment can provide a suitable basis for helping to establish dose constraints, where these are appropriate (see regulation 9(4)). Employers can take account of information about past operating experience and recommendations from relevant professional bodies or trade associations in establishing dose constraints.

Radiation risk assessment for simple uses of ionising radiation

75 The radiation risk assessment is a tool to help employers decide on the most appropriate control measures for new work activities. Completing the assessment is a straightforward process for employers involved in simple uses of ionising radiation, for example an X-ray machine in a mail room. In these cases, the radiation protection issues are not likely to be complex. Often, the assessment will require nothing more than a judgement based on the advice provided by the manufacturer or supplier of the device intended for use. In particular, the assessment will be straightforward where:

(a)   employers follow the advice of the RPA; or

(b)   employers decide to adopt controls and working procedures which adhere to industry standards or are contained in codes of practice or guidance; or

(c)   the work involves the use of a piece of equipment which has been designed to keep exposures as low as reasonably practicable, and which is functioning properly.

76   Advice on the safe use and maintenance of equipment should be available from the manufacturer or supplier.

Recording the results of the radiation risk assessment

77   Employers with five or more employees must record the significant findings of their general risk assessments required under the Management Regulations. The radiation risk assessment must be similarly recorded. This record, which may be electronic, should be an effective statement of the risks the work presents and will make sure management take the necessary actions to protect employees and others exposed to ionising radiation. Where appropriate, assessment records could be linked with other health and safety records such as:

(a)   the health and safety arrangements required by regulation 5 of the Management Regulations;

(b)   the written health and safety policy statement required by section 2(3) of HSWA;

(c)   any local rules required by regulation 18(1) of IRR17.

Review and revision of the prior risk assessment

78   Workplaces, processes and workers change over time. To make sure that the risk assessment remains suitable and sufficient, regulation 3(3) of the Management Regulations requires employers to review the assessment if there is reason to suspect it is no longer valid or there has been a significant change in the work activity. One example would be the need to consider any necessary changes to the working conditions of a pregnant worker. Employers should decide the frequency of such reviews by taking account of the nature of the work, the degree of risk and the extent of any likely change in the work activity.

79   A significant change in the work activity may include:

(a)   the introduction of a radioactive source of a much larger activity or a source which emits a different type or quality of radiation;

(b)   the use of electrical equipment which produces X-rays of much higher energy;

(c)   the introduction of unsealed sources in an area where only sealed sources have previously been used;

(d)   plant modification, including alterations to engineering controls and safety features;

(e)   changes to the process or methods of work;

(f)   human factors, eg arising from staff turnover.

80   One way in which employers could discover that an assessment is invalid is through checking the results of personal dosimetry or area monitoring. These results could indicate a breakdown of controls and so highlight the need for a formal review of whether the procedures in place are satisfactory.

81   Further advice on the review of risk assessments is available at www.hse.gov.uk/risk/review-your-assessment.htm.

Guidance 8(1)–(4)

Assessments for accident hazards

82   Employers must assess the work with ionising radiation they intend to undertake to identify any reasonably foreseeable radiation accidents (defined in regulation 2(1)).

83   These assessments must take account of the consequences, not only of possible plant and equipment failures, but also of a breakdown in work systems and of unauthorised behaviour at work. The level of detail in the assessment should be proportionate to the circumstances. If a particular accident scenario is shown to be extremely unlikely or trivial in its consequences, the assessment needs to go no further. More information is available at www.hse.gov.uk/humanfactors/resources/risk-assessment.htm.

84   Once the assessment has identified how an accident could occur, regulation 8(3) requires reasonably practicable measures either to prevent it happening, or to limit its consequences. These measures need to remain in place to achieve a continued reduction of risk and are different from the mitigation measures once an accident has occurred, which are likely to be reflected in the contingency plan required by regulation 13. They flow naturally from the analysis of accident causation in the assessment.

Regulation 9 Restriction of exposure

Regulation 9(1)

(1)   Every employer must, in relation to any work with ionising radiation that it undertakes, take all necessary steps to restrict so far as is reasonably practicable the extent to which its employees and other persons are exposed to ionising radiation.

ACOP 9(1)

85   Dose-sharing should not be used as a primary means of keeping exposures below the dose limits.

86   Employers should take particular steps to restrict the exposure of any employees who would not normally be exposed to ionising radiation in the course of their work. The dose control measures should make it unlikely that such people would receive an effective dose greater than 1 mSv per year, or an equivalent dose which exceeds that specified as a dose limit for any other person in Schedule 3.

Guidance 9(1)

Responsibility for restricting exposure

87   Employers working with ionising radiation, as defined by regulation 2(1), have overall responsibility for restricting exposure. They must co-operate with any other employers whose employees are affected by their work with ionising radiation, as required by regulation 16 (and paragraph 293 about outside workers).

General approach to restricting exposure

88   This general duty to restrict exposure covers both the intention to work with ionising radiation and the selection of any source of ionising radiation for the work. One of the first considerations is to decide whether the risk can be avoided by choosing an alternative technique which does not involve ionising radiation. Any alternative techniques chosen which avoid work with ionising radiation should not lead to an overall increase in the health and safety risks that employees and other people will be exposed to.

89   Employers should give priority to improving engineering controls and adopting other means of restricting exposure, including changing the methods of work. However, if a choice has to be made between restricting doses to individuals and

restricting doses to a group of people, priority should be given to keeping individual doses as far below dose limits as is reasonably practicable.

90 In general, the lower the activity of any radioactive source used, commensurate with the work that needs to be done, the easier it will be to make sure that exposure of employees and others is adequately restricted. Similar consideration can be given to the minimisation of unnecessary radiation when X-ray sets are used. This may involve the use of appropriate beam filtration, effective collimation of the X-ray beam and a careful choice of the operating voltage and tube current.

91 Where the design of new facilities is being considered for work with ionising radiation, the employer must consider the construction, commissioning and operation of the facility together with its maintenance and decommissioning to ensure that exposure will be restricted as far as reasonably practicable during the life-span of the facility.

92 Valuable feedback can be gathered about the effectiveness of control measures to restrict exposure by analysing monitoring data collected in accordance with regulations 19(8), 20 and 22. Regulation 20(1) specifically requires that the working conditions in controlled and supervised areas are kept under review. Employers must think about how best to review the available information (and how often) to identify any further action that may be required. In some situations, the information should be supplemented with task-specific dose measurements, for example where doses to individuals might be a significant fraction of a relevant dose limit. A suitable RPA appointed under regulation 14 will be able to advise the employer about collecting and analysing relevant information. Appointed safety representatives (and, where appropriate, an established safety committee or dose reduction group) should also to be consulted about reviews of the data.

93 If there are significant variations in doses for the same individual(s) between different periods in the year, or between different individuals doing similar work, a more detailed review should be carried out. Where the dose for an individual exceeds the investigation level established under regulation 9(8), the employer of that person must carry out a formal investigation (see paragraph 178).

Examination of plans and acceptance testing

94 Where new equipment or apparatus is provided or installed, the employer should arrange for acceptance testing of that equipment or apparatus to make sure that it conforms to specification. The RPA must be consulted about the prior examination of plans and acceptance testing of new or modified sources of ionising radiation and the effectiveness of control measures provided to restrict exposure. Employers should also consult the appointed safety representative(s) about the introduction of new equipment or apparatus.

Restricting exposure of young people

95 Both IRR17 and the Management Regulations require employers to give special consideration to the employment of young people in work with ionising radiation. No young person under the age of 18 can be employed to work with ionising radiation where they would need to be designated as classified (see regulation 21).

96   There are specific dose limits for young people who may be exposed to ionising radiations while undertaking training or studying (see regulation 12 or Schedule 3). Employers must take into account risks which could result from young people's lack of experience or risk awareness. Regulation 19 of the Management Regulations places particular requirements on employers of young people regarding matters such as the risk of accidents. Where the control measures are not sufficient to prevent a significant risk to young people, they should only carry out the work if:

(a)   it is necessary for their training;
(b)   they are supervised by a competent person;
(c)   the risk is reduced to the lowest level that is reasonably practicable.

Application to medical exposures

97   This regulation does not apply to the protection of those undergoing a medical exposure, which is covered by the Ionising Radiation (Medical Exposure) Regulations 2017[6] and regulated by the Department of Health and Social Care and the devolved administrations. However, it does apply to staff that carry out those exposures and to members of the public.

Protection for the skin of the hand

98   Radioactive materials, including those in sealed sources, should not be directly held or manipulated in the hand (or close to the hand) if it is practicable for the task to be completed by other means. The only exception to this is where the skin of the hand is unlikely to receive a significant dose and the employee is unlikely to become significantly contaminated with radioactive substances.

99   For research and other laboratories, small quantities of radioactive substances should only be handled where there are no practicable alternatives. In general, it would be practicable to use local shielding and protective gloves in these circumstances so that external radiation and skin contamination risks are effectively controlled. Although it is impracticable to avoid handling a syringe containing a radiopharmaceutical for the injection of a patient, providing a syringe shield will restrict exposure to external radiation.

100 It should always be practicable to avoid handling or working close to a sealed source used, for example in industrial radiography.

(2)   Without prejudice to the generality of paragraph (1), an employer in relation to any work with ionising radiation that it undertakes must –

(a)   so far as is reasonably practicable achieve the restriction of exposure to ionising radiation required under paragraph (1) by means of engineering controls, design features and by the provision and use of safety features and warning devices;
(b)   provide such systems of work as will, so far as is reasonably practicable, restrict the exposure to ionising radiation of employees and other persons; and
(c)   where it is reasonably practicable to further restrict exposure to ionising radiation by means of personal protective equipment, provide employees or other persons with adequate and suitable personal protective equipment (including respiratory protective equipment) unless the use of personal protective equipment of a particular kind is not appropriate having regard to the nature of the work or the circumstances of the particular case.

Hierarchy of control measures

101   Regulation 9(2) sets out a hierarchy of control measures for restricting exposure. Firstly, in any work with ionising radiation, employers should take action to control the doses received by their employees and other people by means of engineering controls. Only after these have been applied should they consider using supporting systems of work. Lastly, employers should provide PPE to further restrict exposure where this is necessary and reasonably practicable.

102   Establishing control measures at an early stage will help employers to effectively restrict exposure, for example when the facility or device is being planned and designed. This means that the dose control mechanisms can be incorporated into the construction of the facility.

103   Employers must put procedures in place to ensure employees use control measures such as systems of work and items of PPE provided to comply with regulation 9(2). They should typically include:

(a)   checks at appropriate intervals to ensure that these control measures are being properly used;

(b)   prompt remedial action where the control measures break down.

Types of physical control measures

104   Engineering controls and design features are normally built into the facility or device; they include all aspects of the design and construction which restrict exposure. Where appropriate, these controls will be central to the operation of the facility or device – for example, constructing suitable containment and shielding of sources, and designing safety-related control systems ensuring that radiation sources are only accessible where necessary for the work being done. In other cases, engineering controls and design features will be put in place specifically to allow the safe use of the device. Examples include local shielding to reduce emitted radiation and local containment, ventilation or other steps aimed at minimising contamination during work with unsealed radioactive materials.

105   Safety features are intended to help ensure the safe use of the equipment in normal operation and to prevent unintended exposure in the event of a failure of control devices or systems of work. Examples include locks on exposure controls, search and lock-up systems, door interlocks for enclosures, and devices which will terminate an exposure in an emergency.

106   Warning devices indicate the status of the equipment in normal operation and alert operators to faults or failures which have occurred and reduce the safety integrity of the installation, but they will not prevent exposure. Examples include pre-exposure and exposure signals and external radiation or contamination alarms. For some work activities, such as industrial radiography, it is appropriate to supply employees with personal electronic dosemeters fitted with an alarm to alert wearers of a high dose rate, for example when a source has failed to retract into a safe position.

Enclosure and radiation shielding

107   Where reasonably practicable, work involving exposure to external radiation should be carried out in a room, enclosure, cabinet or purpose-made structure which is provided with adequate shielding. In situations where this isn't possible, adequate local shielding should be used. Shielding, including beam collimation, will normally be adequate if designed to reduce dose rates below 7.5 µSv per hour in specific locations where people will be working. For use in public areas or where there is continuous access to the working area by

ACOP	**9(2)**

employees or other people not directly involved in the work, the shielding should be designed to reduce dose rates to the lowest level that is reasonably practicable. In this case, the dose rate should be so low that it is unnecessary to designate the area around the device as a supervised area.

Guidance	**9(2)**

108 In many cases, shielding will either form part of the equipment (eg covers, shutters and collimators) or an enclosure around the device (eg a room or purpose-made structure). Local shielding around sources including purpose-made covers, drapes, free-standing screens and even bags of lead shot can also be used to restrict exposure where an enclosure is not reasonably practicable.

ACOP	**9(2)**

Work with unsealed materials

109 Employers should give priority to the containment of radioactive sources and substances to prevent dispersal or contamination. Where such containment alone is not sufficient to give the required protection, ventilation should be provided.

Guidance	**9(2)**

110 A building, room or enclosure used for work with unsealed radioactive material should incorporate design features which take into account the risk of contamination likely to arise from the work. In particular, employers must make sure surfaces are easy to decontaminate and clean. They must also make arrangements to safely decommission or dismantle equipment that may have become internally contaminated.

111 Employers should choose fume cupboards, sterile cabinets, glove boxes, ducting, fan assemblies, filtration units and other components of the ventilation system that have been designed and constructed specifically for radioactively contaminated atmospheres. Equipment must be designed and constructed to make it easy to maintain, clean and decontaminate. Disposal of filter media and other such contaminated parts of the system should be possible with minimum personal exposure to radiation. The design and operation of the ventilation system will need to satisfy any subsequent environmental regulations and are required to minimise any waste arising from that work.

112 The provision and cleaning of washing facilities and changing facilities are dealt with under regulation 19(10).

113 High levels of radon and its decay products can be reduced in buildings by use of continuous engineering controls, such as a radon sump and pump system or positive ventilation. Use of ventilation and other engineering controls is also an effective way of reducing levels of radon in mines and other facilities. Ventilation engineering expertise may be needed to optimise protection.

ACOP	**9(2)**

Exposure controls

114 Where control systems permit, interlocks or trapped key systems should be provided and properly used where they can prevent access to high-dose-rate enclosures. They should be fitted so that the control system will ensure an exposure:

    (a)   cannot start while the access door, access hatch, cover or appropriate barrier to the enclosure is open;

    (b)   is interrupted if the access door, access hatch, cover or barrier is opened;

    (c)   does not restart solely when closing a door, access hatch, cover or barrier.

115   It should always be reasonably practicable to design control units for X-ray generators and radioactive source containers to prevent unintended and accidental exposure. To avoid any possible confusion, control switches should be labelled clearly.

116   Employers must consider how an exposure will be stopped or a radioactive source returned to its shielding if the normal control system malfunctions.

117   On many inspection and gauging devices the material being examined is taken past the imaging or sensing system. There will be no need for a radiation beam if the material transfer system is not operating. It is normally appropriate for the exposure control system to terminate an exposure or close the shielding shutter when the transfer system stops or is stopped.

118   For high-dose-rate radiation enclosures, employers must make sure that no employee remains inside when exposures begin. Examples of such facilities include shielded enclosures used for industrial radiography, medical radiotherapy suites and gamma and electron beam sterilisation facilities. Facilities in which it may be possible to exceed a relevant limit on equivalent dose to part of the body (eg to the hand) within a few minutes would include enclosures around X-ray optical equipment.

119   Robust systems of work should prevent the possibility of people being trapped in a high-dose-rate enclosure (see paragraphs 122–129). The consequences of a person being trapped in such an enclosure are potentially very serious. A radiation risk assessment under regulation 8 may indicate a need for further action; for example it will be appropriate to place emergency devices such as 'off' buttons in appropriate locations to allow a person to prevent or quickly interrupt the emission of ionising radiation.

120   The ACOP advice in paragraph 114 means that effective interlock devices must normally be designed and installed in such a manner that if they fail to operate correctly no exposure can occur.

121   Where there is a risk of receiving an exposure within a few minutes that could cause deterministic effects, it should be reasonably practicable for employers to install engineered search and lock-up systems into the exposure control system. These systems require the full area of the enclosure to be properly checked and vacated before they can be initiated. The check would be confirmed by the operation of 'search' buttons within a predetermined time before the area is closed and irradiation begins.

122   If, despite the use of exposure controls, employees or other people could become trapped in a high-radiation-dose enclosure, for example an industrial radiography enclosure, employers should install a suitable alarm system. This will indicate that an exposure is imminent. Employees will need access to an 'emergency' control system which will enable them to prevent the initiation of the exposure or to stop the exposure themselves. Otherwise, they should be able to alert the operator(s) outside the enclosure to their presence and to take refuge in a shielded part of the enclosure without passing through the main beam. The dose rate in that shielded refuge should preferably be below 7.5 µSv per hour during any exposure.

Key-operated safety devices

123   Where there is a risk of significant exposure arising from unauthorised or malicious operation of X-ray generators or radioactive source containers, employers should use equipment which has been fitted with locking-off arrangements to prevent its uncontrolled use.

ACOP   **9(2)**

124 The initiation of exposures should be under key control, or by some equally effective means, to prevent unintended or accidental emission of a radiation beam or exposure of a source. This is particularly important where the control point is remote from the equipment which will be activated or there is general access to equipment by members of the public or personnel who are not carrying out the work with ionising radiation.

Guidance   **9(2)**

125 Where key-operated devices are provided, employers must make sure that keys are only available to authorised employees. Trapped key systems will help to prevent unauthorised or accidental operation of the equipment.

ACOP   **9(2)**

Warning devices

126 Sources of ionising radiation which can cause significant exposure in a very short time should be fitted with suitable warning devices which:

(a) indicate for a radioactive source whether it is in or out of its shielding (or the exposure shutter is open or closed);

(b) indicate for an X-ray generator when the tube is ready to emit radiation and, except for diagnostic radiology, give a signal when the useful beam is about to be emitted and a distinguishable signal when the emission is under way unless this is impracticable;

(c) for X-ray generators other than those used for diagnostic radiology, are designed to be automatic and fail-safe, ie if the warning device itself fails the exposure will not proceed.

Guidance   **9(2)**

127 For most X-ray generators and some sealed sources it should be reasonably practicable to have automatic warning devices. Employers using large sealed sources must consider how signals could be given. In particular, automatic generation of the pre-exposure and exposure warning signals must be possible where there is a fully automated mechanism for the exposure and retraction of the source.

ACOP   **9(2)**

128 Employers should make sure that warning signals can be seen or heard by all those people who need to know the status of the source of ionising radiation for protection purposes.

Guidance   **9(2)**

129 This includes people outside the immediate area. The employer should provide explanatory notices for people not directly involved in the work, pointing out any action they should take in response to a warning signal. These notices will only be useful if they explain clearly the significance of the different signals and the action to be taken. Employers should consult the appointed safety representative(s) for the area about such notices.

Manual warning signals, radiation monitors and warning labels

130 Where equipment containing a radioactive source is not fitted with automatic warning signals, the operator must use a manual system to generate clear and distinguishable pre-exposure and exposure warning signals, eg the use of portable gamma alarms in site radiography. Signals are particularly important when people in or around the designated area need to be aware of the status of the source. Notices and signs can explain the significance of these signals and the action to be taken. Appointed safety representatives should be consulted on manual warning systems.

Systems of work

131 The engineered safety features should, so far as reasonably practicable, be supported by systems of work, incorporated into the local rules, to be followed by

employees and other people when in the vicinity of a source of ionising radiation. These systems of work could include 'permit-to-work systems' that allow strict management control over the conditions in which work will proceed, how it will be done and how it will be supervised.

132 Employers must consult the appointed safety representative(s) relevant to the area about any proposal to introduce a permit-to-work system.

133 Employers must take simple steps to control the exposure of people, for example by making sure that:

(a) they control the number of people present in high-dose-rate or contaminated areas and the time they remain there is limited;

(b) in interventional radiology, staff not required to be close to the couch and involved in the direct clinical examination or care of the patient remain away from the high-dose-rate areas (which requires a good understanding of dose-rate contours around the couch) and preferably be behind a shielded screen;

(c) in a storeroom containing a large number of different radioactive sources, any areas or zones in which the dose rate is significantly high are well known and work in the storeroom is organised so that the time spent in these areas or zones is minimised, taking account of the impact on external dose rates outside the storeroom;

(d) for transport of radioactive material by road, the system of loading is such that packages emitting a higher dose rate are positioned furthest from the driver's seat, reducing the driver's exposure;

(e) work with unsealed radioactive substances is carried out in a fume cupboard and lipped trays with absorbent lining are used to contain minor spillages.

134 Signs used to indicate high-dose-rate areas must conform to the Health and Safety (Safety Signs and Signals) Regulations 1996.[7] Where the area concerned has been designated as a controlled area or a supervised area, the requirements of regulation 18 will also apply.

135 It is vital to produce detailed safe systems of work for equipment in which any of the engineered safety systems have been legitimately disabled, eg as part of special maintenance procedures. Close supervision of the work will be necessary. Examples include:

(a) the loading or unloading of sealed sources in or from equipment where normal safeguards for restricting exposure cannot be followed;

(b) the setting up and alignment of X-ray optics equipment;

(c) maintenance work on a thickness gauge incorporating a sealed source which involves any work inside the gauge head or with the gauge heads split.

136 Employers should provide systems for cleaning up minor spillage of radioactive material during normal work. If it is not reasonably practicable to clean up all the contamination in an area or on a piece of work equipment, special restrictions might then need to be applied to future work in that area. If the contamination consists of short-lived radionuclides, it may be better to allow these to decay rather than to take immediate action to clean up the area. There will be circumstances where it is necessary to clean up radioactive contamination even at the expense of increasing the exposure of a group of employees involved in the clean-up, especially where employees who are not normally exposed or the public have access to that area.

Guidance	**9(2)**

137 Employees must be trained to follow systems of work and to interpret and respond to any warning signals or supporting notices which they may encounter in the course of their work.

ACOP	**9(2)**

138 Employers working with ionising radiation should make sure that a check is made with a suitable radiation monitoring instrument after each exposure using high-dose-rate sealed source equipment (such as that generally used for industrial radiography or processing of products). The purpose of the check is to establish that the sealed source has fully retracted to its shielded position and that the area is safe to enter. In addition, employees engaged in this type of work should wear a dosemeter which gives an audible alarm when high dose rates are detected.

Guidance	**9(2)**

Monitoring radiation levels as part of a system of work

139 As well as providing monitoring instruments for routine work with ionising radiation (see also regulation 20), employers will need to ensure that such instruments are used as part of the system of work for certain non-routine operations such as maintenance work on high-dose-rate sealed source equipment. Employees involved in this type of work should wear personal alarming dosemeters for such operations, in addition to any other dosimetry provided for dose recording.

Adequate support for employees and others

140 Employers must decide on the number of trained people needed to allow a job to proceed safely; for example in site radiography work the radiographer will normally require the support of at least one other person. This assistant will usually need to patrol the boundaries of the controlled area to ensure that access is restricted (see regulation 19) and perform the necessary radiation monitoring of that area (see regulation 20) and would be available to assist with implementing the contingency plan (see regulation 13).

Regulation	**9(3)**

(3) An employer who provides any system of work or personal protective equipment pursuant to this regulation must take all reasonable steps to ensure that it is properly used or applied as the case may be.

Guidance	**9(3)**

141 Personal protective equipment (PPE)[8] includes respiratory protective equipment (RPE),[9] protective clothing, footwear and equipment to protect the eyes. Types of PPE specific to protection against external ionising radiation include protective aprons, gloves and eyewear. Various types of RPE (including pressurised suits) minimise the risk of inhaling radioactive material.

142 The term 'adequate' in regulation 9(2)(c) refers to the ability of the equipment to protect the wearer. The term 'suitable' refers to the correct matching of the equipment to the job and the person. To be considered 'adequate and suitable' PPE should be correctly selected and used.

143 The risk assessment should be used to decide on the choice of PPE (see paragraph 71(c)). The purpose of the assessment is to make sure that the employer chooses PPE which is adequate and suitable. For RPE this implies that it provides an adequate margin of safety and is matched to the job, the environment, the anticipated air concentration of radioactive material and the wearer.

144 The performance of RPE which relies on a tight-fitting facepiece will be adversely affected if there is not good contact between the wearer's skin and the face seal of the mask, for example because the wearer is not clean-shaven. RPE from more than one manufacturer and of more than one type may be needed to meet the face-fit requirements of all the employees in a particular area. Proper

Guidance 9(3)

testing of the correct fit for facepieces intended to fit tightly is a necessary part of the selection process. For more information on the selection, use and maintenance of suitable RPE, see HSE's website: www.hse.gov.uk/respiratory-protective-equipment/.

145 Sometimes, PPE could introduce other forms of risk greater than that arising from the ionising radiation. In these circumstances, it might not be appropriate to use PPE. Examples might include use of bulky breathing apparatus in areas with limited space for free movement, or an employee being unable to carry the weight of a lead apron without risk of injury.

146 HSE guidance on the Personal Protective Equipment at Work Regulations 2002[10] provides advice about the selection and use of PPE which is equally applicable to protection against ionising radiation.

Regulation 9(4)–(5)

(4) Where it is appropriate to do so at the planning stage of radiation protection, an employer, in relation to any work with ionising radiation that it undertakes, must use dose constraints in restricting exposure to ionising radiation pursuant to paragraph (1).

(5) An employer must establish the dose constraints referred to in paragraph (4) in terms of individual effective or equivalent doses over a defined appropriate time period.

Guidance 9(4)–(5)

What are dose constraints?

147 A dose constraint is an upper level of individual dose specified by the employer. It must be set out at the planning stage of the work. It is one of many tools for helping to restrict individual exposures so far as is reasonably practicable. Dose constraints may be used to consider the best plan or design for an individual task or event or the introduction of a new facility. However, they are not intended to be used as investigation levels once a decision has been taken about the most appropriate design or plan (see paragraph 180).

148 The value of a dose constraint represents a level of dose (or some other measurable quantity) which should not be exceeded in a well-managed workplace.

149 Realistic predictions of individual doses associated with any proposed control measures for restricting exposure should be compared with the value selected for the dose constraint. If the predicted doses exceed the value of the dose constraint, the employer would be expected to choose better control measures. These should then lead to predicted doses below the dose constraint. Therefore, a dose constraint should help to filter out options for radiation protection that could lead to unreasonably high levels of individual dose, even though the collective dose for the workforce as a whole is optimised.

150 A dose constraint is not a maximum dose. In some cases, the employer may decide that it is acceptable for predicted doses to exceed the dose constraint where, for example, other health and safety risks have to be taken into account in selecting the most appropriate control measures.

When is it appropriate to use dose constraints?

151 Dose constraints are generally appropriate for members of the public who may be affected by direct radiation or contamination arising from work activities, not least because those individuals may be exposed to more than one such source. However, for occupational exposure, dose constraints may only be appropriate in a limited number of situations.

152 The need to establish dose constraints should be considered as part of the risk assessment process (see regulation 8). Where the employer establishes one or more dose constraints it would be sensible to record these as part of the arrangements made under the written safety policy required by HSWA.

153 Dose constraints are meant to represent levels which are normally achievable in the particular type of work activity in well-managed operations using effective physical controls and systems of work to restrict exposure. Where an employer decides it is appropriate to establish dose constraints for particular types of work, these could be based on past operating experience and on any recommendations from relevant professional bodies or trade associations.

Dose constraints for outside workers

154 Dose constraints for outside workers should be established in co-operation with the employer of the outside worker and the undertaking where the outside worker will be working with ionising radiation.

Dose constraints for members of the public

155 Where employers anticipate that any work activity or facility is likely to expose members of the public to direct radiation or contamination, they should apply a dose constraint. It is recommended that the constraint on optimisation for a single new source should not exceed 0.3 mSv a year. Employers should take this recommendation into account in establishing a dose constraint for members of the public. The constraint should be applied to estimates of dose for representative individuals likely to receive the highest average dose from the work.

156 Employers in the healthcare sector should consider the use of dose constraints for members of the public who are not carers and comforters but are likely to receive some exposure as a result of sharing transport or accommodation with a patient who has received a therapeutic administration of a radiopharmaceutical. However, employers may decide that sufficient action has been taken by following accepted practice on the release of such patients from hospital.

Dose constraints for occupational exposure

157 Dose constraints for occupational exposure are only likely to be appropriate where individual doses from a single type of radiation source will be a significant fraction of the dose limit (ie at the rate of a few mSv a year). Dose constraints are not likely to be appropriate for occupational exposures resulting from the following types of work with ionising radiation where employee doses tend to be low:

(a)   diagnostic radiology, nuclear medicine, most radiotherapy and other medical exposures;
(b)   most work in the non-nuclear industrial sectors;
(c)   teaching and most research activities.

158 The main exception would be for special types of work (eg some interventional radiology), where effective doses are likely to be more than a few mSv per year.

159 Even in specialised areas, such as industrial radiography, where individual doses are sometimes relatively high, it may not be appropriate to establish dose constraints for planning individual jobs unless adequate dose information is available for that type of work. If the employer has such information, it should be

possible to choose a dose constraint which is representative of a well-managed operation, for example radiography of steam tubes during a conventional power station outage.

160 The nuclear sector has considerable experience in developing dose databases relating to good practice. Therefore, dose constraints may be appropriate for some well-defined operations in this sector. Any dose constraints established by such employers should be consistent with ONR's Safety Assessment Principles.[11]

Regulation   9(6)

(6)   Without prejudice to paragraph (1), an employer who undertakes work with ionising radiation must ensure that –

(a)   in relation to an employee who is pregnant, the conditions of exposure are such that, after the employee's employer has been notified of the pregnancy, the equivalent dose to the foetus is as low as is reasonably practicable and is unlikely to exceed 1 mSv during the remainder of the pregnancy; and

(b)   in relation to an employee who is breastfeeding, that employee must not be engaged in any work involving a significant risk of intake of radionuclides or of bodily contamination.

Guidance   9(6)

Assessing risks for pregnant or breastfeeding employees

161 Nothing within these Regulations prevents pregnant or breastfeeding employees from working with ionising radiation provided that exposures remain below the dose limit specified and are kept as low as is reasonably practicable.

162 The general risk assessment required by regulations 3 and 16 of the Management Regulations must take account of risks to the health and safety of a new or expectant parent at work, or to that of the baby or foetus. For any new activities, the radiation risk assessment required by regulation 8 of IRR17 should indicate to the employer what exposures pregnant or breastfeeding employees are likely to receive in particular working areas. These assessments will identify areas where the foetus could receive a dose greater than 1 mSv during the declared period of pregnancy. The RPA should be involved in that assessment.

163 For exposure to external radiation, this dose restriction is broadly equivalent to a dose to the surface of the abdomen of a pregnant employee of about 2 mSv in many working situations, for example exposure arising from the diagnostic use of X-rays in hospitals. Where exposure is to high-energy radiation, employers will need advice from the RPA about an appropriate dose restriction. If the pregnant or breastfeeding employee is likely to receive significant intakes of radionuclides as a result of working conditions in an area (typically where the committed effective dose is of the order of 1 mSv a year or more), particular care may be needed. In these cases the dose to the foetus may approach or exceed 1 mSv for certain radionuclides which are absorbed more readily by the tissues of the placenta and foetus.

164 Employers must also consider risks relating to employees who are breastfeeding. The employer's risk assessment should show whether there are any working areas where a worker may receive significant bodily contamination which could be a risk to them or their child. They can take account of any control measures provided to prevent such contamination, including the use of PPE, provided that effective steps have been taken to make sure it is used at all times. In deciding whether bodily contamination may be significant, employers should be aware that certain radionuclides are likely to become concentrated in breast milk and that the dose to the infant may be of much greater concern than the dose to the breastfeeding employee.

Declaration of pregnancy

165 Control measures can only be put in place if the employee declares their pregnancy to the employer. Employees must understand the importance of notifying the employer in writing, even if the employee would prefer to keep their condition completely confidential. Employers must inform their employees about the possible risks and the importance of notifying the employer in writing as soon as the employee knows that they are pregnant.

166 Employers may wish to arrange for employees to receive counselling from the relevant doctor or an occupational health service to discuss the possible risks involved. These arrangements might also provide an opportunity for an employee who becomes pregnant to discuss their condition with a health professional and also to tell their employer about the pregnancy accordingly.

167 If the employee informs the employer of the pregnancy for the purpose of any other statutory requirements, such as maternity pay, this will be sufficient in the absence of any earlier notification, and the employer should act on the information for the purposes of this regulation.

Return from maternity leave

168 When an employee returns from maternity leave to work involving unsealed sources and the employer's assessment shows that bodily contamination is reasonably foreseeable, it is advisable to assume that an employee may be breastfeeding and to take appropriate action, notwithstanding the provision in regulation 9(6). The employer's action should include giving advice about the possible risks (regulation 15) and altering the employee's working conditions. The relevant doctor or occupational health service may provide counselling if appropriate. It is sensible for employees continuing to breastfeed for more than six months to notify the employer to make sure that any special measures to prevent bodily contamination can continue.

Need for co-operation between employers

169 Employers using contractors to carry out any work with ionising radiation must make arrangements with the contractor's employer (under regulation 16) to be informed if any of the employees:

(a) have declared themselves to be pregnant; or
(b) in cases where bodily contamination is reasonably foreseeable, have declared that they are breastfeeding.

170 Both employers have responsibilities under health and safety law to assess health and safety risks at work.

171 This does not apply if the pregnant or breastfeeding employee, the foetus or the infant would not be at risk from ionising radiation during that work.

172 Similar responsibilities apply to self-employed contractors who may be pregnant or breastfeeding.

Action required by employers working with ionising radiation

173 After being informed of a pregnancy, employers working with radiation must decide if any restrictions are necessary in the particular case, taking account of the results of the risk assessment.

174   If the assessment shows some action is necessary, this may be limited to avoiding work:

(a)   with large sources of external radiation where there is a reasonably foreseeable risk that pregnant employees may receive a significant accidental exposure; or

(b)   where there is a significant risk from intakes of radionuclides.

175   If such risks cannot be avoided in line with regulation 9(6) of IRR17, regulation 16 of the Management Regulations requires the employer to:

(a)   alter the hours of work of the employee or, where this would not be reasonable;

(b)   identify and offer the employee suitable alternative work that is available, or where this is not feasible;

(c)   suspend the employee from work on full pay.

176   The Employment Rights Act 1996[12] requires that this suspension should be on full pay. Employment rights are enforced through the employment tribunals.

177   Further guidance on new and expectant parents is available in HSE's publication *New and expectant mothers who work.*[13]

(7)   Nothing in paragraph (6) requires the employer who undertakes work with ionising radiation to take any action in relation to an employee until that employee's employer has been notified in writing by the employee of the pregnancy or that the employee is breastfeeding and the employer who is undertaking the work with ionising radiation has been made aware, or should reasonably have been expected to be aware, of that notification.

(8)   Every employer must, for the purpose of determining whether the requirements of paragraph (1) are being met, ensure that an investigation is carried out without delay when the effective dose of ionising radiation received by any of their employees for the first time in any calendar year exceeds 15 mSv or such other lower effective dose as the employer may specify, which dose must be specified in writing in local rules made pursuant to regulation 18(1) or, where local rules are not required, by other suitable means.

Formal investigation levels

178   The purpose of this regulation is to trigger a review of working conditions when an employee (or a group of employees carrying out similar work) have recorded doses which exceed the specified investigation level, to make sure that exposure is being restricted so far as is reasonably practicable. The investigation has a different purpose from that required by regulation 26 concerning possible overexposures.

179   The duty to carry out an investigation is placed on the employer of the person whose recorded dose has exceeded the investigation level. In most cases, this will probably be an employer who is working with ionising radiation. However, where the employer is that of a contractor (eg a scaffolding contractor or a cleaning company) working on one or more sites belonging to other employers who work with ionising radiation, the investigation may have to take account of work with ionising radiation carried out at all these different sites throughout the calendar year.

180   The regulation requires employers to specify an appropriate investigation level for relevant groups of employees which will normally be significantly less than 15 mSv a year. Employers should take into account the profile of doses for employees as a

whole (or particular groups) and seek advice from the RPA before selecting an appropriate investigation level. The appointed safety representative(s) or the established safety committee should be consulted about the choice of the investigation level. Employers cannot specify an investigation level higher than 15 mSv a year.

181 Employers should select different investigation levels for different sites or different groups of employees, where appropriate.

182 Employers should make arrangements for reviewing doses reported in dose summaries for classified persons by an approved dosimetry service (see regulation 22(3)(b)) or for other people entering controlled areas (see regulation 19(5)). Such arrangements will provide an early warning if an employee's cumulative dose for the year is approaching the investigation level. In these cases, the employer should consider if it is appropriate to take further measures to restrict exposure before a formal investigation becomes necessary.

183 A formal investigation under regulation 9(8) should aim to check the adequacy of the control measures provided to restrict exposure, to see if further engineering controls etc are appropriate. It should include such matters as:

(a) details of the work routine of the individual and immediate work colleagues for the year so far;
(b) evidence about the involvement of the individual in any known incidents in which they may have received an unusual exposure;
(c) details of that person's estimated exposures by task or relevant dose assessment period, compared with estimated exposures of work colleagues undertaking similar work;
(d) results of any special radiation survey in the areas where the person worked, to identify any deterioration in physical control measures;
(e) evidence from the radiation protection supervisor (RPS), the individual concerned and work colleagues about adherence to local rules or deficiencies in those rules in the light of changes to work practices.

184 An effective investigation should produce firm conclusions about the need for further control measures (or the better application of current controls). Those conclusions would normally be recorded together with the main findings in an investigation report. If those conducting the investigation conclude that further action is appropriate they should consider making recommendations to senior management, as appropriate. That action might include changes to physical control measures or systems of work and planning exposures for the individual or group for the remainder of the year.

185 As in the ACOP advice in paragraph 249, employers must normally consult the RPA about such investigations.

186 Where groups of employees are engaged in similar work in the same type of environment, only one investigation may be needed if two or more individuals receive doses above the investigation level.

187 Employers should consult any appointed safety representative about the formal investigation.

188 Employers need to be able to demonstrate to the appropriate authority that an adequate investigation has been carried out. Retaining a copy of the investigation report for at least two years is sufficient for this purpose.

Regulation 10 Personal protective equipment

Regulation 10(1)–(2)

(1) Any personal protective equipment provided by an employer pursuant to regulation 9 must be suitable for its purpose and –

(a) comply with any provision of the Personal Protective Equipment Regulations 2002 which is applicable to that item of personal protective equipment; or

(b) in the case of respiratory protective equipment where no provision referred to in sub-paragraph (a) applies, be of a type approved or conform to a standard approved in either case by the Executive.

(2) Every employer who provides personal protective equipment pursuant to regulation 9 must ensure that adequate facilities are provided for the storage of that equipment.

Guidance 10(1)–(2)

189 The results of the radiation risk assessment will determine the need for personal protective equipment (PPE) in accordance with the Personal Protective Equipment Regulations 2002[9] and the Personal Protective Equipment at Work Regulations 1992 (as amended)[8] and should indicate whether PPE will contribute to the reduction of exposure. Use of PPE should be a last resort when all other control measures to reduce the risk of radiation exposure have been considered, such as engineering controls and safe systems of work. Advice on the selection of adequate and suitable PPE for restricting exposure should be obtained from an RPA. Appointed safety representative(s) should also be consulted about the suitability of PPE for the wearer. More information can be found at www.hse.gov.uk/toolbox/ppe.htm and professional associations also produce their own guidance.

190 Employers must make sure that PPE is maintained so that it performs correctly. For more information on maintenance see regulation 11.

Regulation 11 Maintenance and examination of engineering controls etc and personal protective equipment

Regulation 11(1)

(1) An employer who provides any engineering control, design feature, safety feature or warning device to meet the requirements of regulation 9(2)(a) must ensure –

(a) that any such control, feature or device is properly maintained; and

(b) where appropriate, that thorough examinations and tests of such controls, features or devices are carried out at suitable intervals.

Guidance 11(1)

General advice on maintenance of controls

191 'Maintained' is defined under regulation 2 in relation to plant, apparatus, equipment and facilities. 'Maintenance' means 'maintained in an efficient state, in efficient working order and good repair'.

192 This regulation requires that all physical control measures (shielding, enclosure, ventilation, safety features and warning devices etc) provided in accordance with regulation 9(2) are operating correctly, ensuring their condition does not deteriorate so it can no longer provide the required level of protection. In such cases where failures are revealed, for example during the examinations and tests, employers must either:

(a)   stop that work with ionising radiation; or

(b)   take other effective action under regulation 9(2) to restrict exposure;

until the enclosure or device is restored to efficient working order.

193   Employers are likely to be meeting the obligations and requirements of this regulation where:

(a)   they hold a site licence granted under the Nuclear Installations Act 1965;[14] and

(b)   controls, features or devices are already maintained and tested in accordance with relevant conditions in that licence.

Nature of inspections and tests

194   Employers working with ionising radiation should put in place formal programmes for the inspection and test of active design features, such as exhaust ventilation systems. For passive engineering controls and design features, they can use data from monitoring programmes provided under regulation 20. This data can be used to check the effectiveness of radiation shielding and the systems in place for containing and minimising contamination for work with unsealed materials. For example, routine contamination monitoring can be used to check the continuing suitability of the easy-to-clean surface on a table in a laboratory. In this case, employers should make use of the monitoring information to check that the surface finish is maintained to the required standard.

195   The effectiveness and performance of the active engineering controls should be assured and checked when any acceptance test is done, before the equipment is first put into use. The tests and inspections done under this regulation can be seen as repeating a sample of these original acceptance tests at suitable intervals, allowing employers to maintain the performance of these engineering controls at the initial level.

196   Employers should carry out a visual check of all active control measures used to prevent or restrict exposure, provided carrying out such checks does not present undue risk to employees. Where appropriate, these checks should be carried out at frequent intervals, for example during every shift for safety-critical features.

197   For the examination to be sufficient employers must make sure that the person carrying out the examination or test is competent.

198   Employees must co-operate in maintaining controls by:

(a)   storing equipment and PPE appropriately;

(b)   reporting visible damage or malfunctions immediately.

Frequency of tests and inspections

199   In addition to the visual checks already mentioned, active control measures such as exhaust ventilation systems, electromechanical or other types of interlocks and warning devices should be examined and tested at suitable intervals.

200   Typically, a suitable interval for the examination and test of active control measures is about once a year, depending on the circumstances of the work. If failure of control measures could result in an acute exposure above the investigation level established under regulation 9(8), the frequency of examination

and test of that control measure should be more than once a year. In deciding on the required test and inspection frequency, employers should consider:

(a) possible dose implications of a failure;
(b) reliability of the control measure;
(c) doses likely to be received by workers while carrying out any examination or test.

Consultation on procedures for maintenance

201 Regulation 14 and Schedule 4 require employers to consult a suitable RPA on the periodic examination and testing of engineering controls etc appropriate to the equipment they use (see regulation 14). Employers should:

(a) obtain RPA advice to develop a programme of examination and testing for appropriate control measures;
(b) consult the appointed safety representative(s) or established safety committee;
(c) follow the recommendations of the manufacturer or supplier on the programme for the inspection and test of the different control measures.

Recording the results

202 Sufficient records should be kept of these examinations and tests to enable employers to identify:

(a) which controls, features or devices have been examined or tested;
(b) what action is required to maintain them; and
(c) when the next examination or test is due.

203 Employers need to keep suitable records of monitoring and tests carried out and retain details such as the date and nature of the test, where it was carried out and by whom, as well as the information in paragraph 202. The details must be held in paper or electronic records and should be available for inspection if required. Records should be kept for two years after the subsequent test. These might, for example, take the form of a general maintenance log for all controls, features and devices provided under regulation 9(2)(a).

(2) Every employer must ensure that –

(a) all personal protective equipment provided pursuant to regulation 9 is, where appropriate, thoroughly examined at suitable intervals and is properly maintained; and
(b) in the case of respiratory protective equipment, a suitable record of that examination is made and kept for at least two years from the date on which the examination was made and that the record includes a statement of the condition of the equipment at the time of the examination.

Maintenance etc of personal protective equipment

204 The purpose of a thorough examination is to establish that the item being examined is fit for the intended use and there has been no physical deterioration since its last thorough examination. Disposable PPE, such as gloves, do not need a thorough examination and test unless they are reused after a working session. In this situation, more frequent checks for possible radioactive contamination are needed (eg after each shift or wear period) to comply with the general duty under regulation 9(2)(c).

205 The interval required between thorough examinations will depend on the:

(a) type of equipment;
(b) hazard against which it is required to protect;
(c) conditions under which it is used;
(d) likelihood of deterioration;
(e) frequency of use.

206 Thorough examination and testing must be carried out by people who are competent to do this work and in accordance with the manufacturer's instructions.

207 For RPE, the interval between each examination and test should not exceed one month. For self-contained breathing apparatus, the examinations should follow the manufacturer's instructions, for example a check on the condition of the air supply. For RPE and ready-to-use PPE kept in sealed packs, as recommended by the manufacturer, check that it has no obvious deterioration and it is still 'in date' for use. HSE publication *Respiratory protective equipment at work: A practical guide* (HSG53)[15] gives further advice.

Records of examination for personal protective equipment

208 Formal records of examination are required for RPE. They should contain the:

(a) means of identification and condition of the equipment;
(b) date of examination and signature of the person who carried it out.

209 Where many similar items of RPE are involved, employers should adopt a dating system, ensuring the date of the last examination is known, or can be worked out, and that defective equipment can be identified for removal from service. Records should be kept for at least two years following the subsequent examination. The records must be held in a format that can be made available for inspection if required.

Regulation 12 Dose limitation

(1) Subject to paragraph (2), every employer must ensure that its employees and other persons within a class specified in Schedule 3 are not exposed to ionising radiation to an extent that any dose limit specified in Part I of that Schedule for such class of person is exceeded in any calendar year.

210 Assessments of effective dose and equivalent dose from external radiation, for the purpose of comparison with the dose limits specified in Schedule 3 of the Regulations, should be made using the operational quantities defined in Appendix 2.

211 Assessments of committed effective dose, and committed equivalent dose following intakes of radionuclides into the body, should take account of the likely dose over 50 years following the intake (up to age 70 for children). They should also be attributed to the calendar year of the intake so they can be compared to dose limits.

Responsibility for ensuring compliance with dose limits

212 The main requirement in these Regulations is set out in regulation 9. Employers must make sure that exposures arising from the work are kept as low as reasonably practicable. Complying with dose limits is an absolute requirement and in most cases during routine exposure it is unlikely that an individual's dose will

approach a relevant dose limit. Doses received by employees who are not normally exposed to ionising radiation in the course of their work should be well below dose limits (see ACOP advice in paragraph 86). However, all employers must make sure that the cumulative exposure of their employees over the year does not exceed a relevant dose limit (see regulation 16 concerning co-operation between employers).

Demonstrating compliance with dose limits

213   The dose limit quantities 'effective dose' and 'equivalent dose' are quantities specified by the International Commission on Radiological Protection. Effective dose relates to the whole body; equivalent dose relates to the dose received by a tissue or part of the body, for example the skin or the eye. Appendix 2 defines these dose quantities and the operational quantities to be used for individual monitoring in relation to external radiation.

214   In regulation 2(6), any reference to 'effective dose' means the sum of the effective dose from external radiation and the committed effective dose from internal radiation, unless the context requires otherwise. 'Equivalent dose' also includes the committed equivalent dose from internal radiation. The 'committed dose' from an intake of a radionuclide is the dose assessed to be received by a tissue or organ over at least a 50-year period following the intake (70-year period for children). The intake might occur from inhalation or ingestion of radioactive contamination or from radionuclides entering wounds such as a cut to the hand.

215   If the biological or physical half-life of a radionuclide is short, all of the dose will probably be received in the calendar year in which the intake was received. However, if the radionuclide is long-lived and is likely to remain in the body for many months or years, the dose is likely to be received over a number of calendar years. It is internationally accepted convention that the 'committed dose' for an employee is assigned to an individual's dose record for the calendar year of the intake.

216   Approved dosimetry services (and often RPAs) have the relevant expertise in dealing with these quantities. Dosimetry services approved to assess doses received by classified persons from external radiation are currently expected, as a condition of approval, to use the operational dose quantity $H_p(d)$ specified in ICRP Publication 116,[16] section 2.3 or personal dose monitoring (see Appendix 2). Those dosimetry services approved to assess committed effective dose (and committed equivalent dose) for classified persons following intakes of radionuclides into the body will generally use the dose coefficients in ICRP publication 119[16] and its updates, where appropriate.

Application of dose limits to different classes of person

217   Schedule 3, Part 1 specifies relevant dose limits for different classes of person – adult employees aged 18 or over, trainees and students aged 16 to 18, and any other person. The limit on effective dose is:

(a)   20 mSv a year for any person aged 18 or over who is an employee (as defined by regulation 2(2)(b) of these Regulations), except as provided by regulation 12(2) and Schedule 3 Part 2. However, a higher effective dose of up to 50 mSv may be authorised by the appropriate authority in a single year, provided that the total dose over any five consecutive years, including the years for which the limit has been exceeded, does not exceed 100 mSv;

(b)   6 mSv a year for trainees aged between 16 and 18 (see definition in regulation 2(1)) and students aged between 16 and 18 who work with ionising radiation in the course of their studies;

(c) 1 mSv a year for any other person, eg members of the public and employees under 18 years of age who are not trainees (but see paragraph 223 about special cases in which doses to members of the public may be averaged over five years).

218 The limits only relate to exposures received by individuals within that class. The exposure of employees while they are at work is considered separately from any exposure they might receive when not at work with ionising radiation. Therefore the limits for adult employees will only apply to occupational exposures. Exposures received by an individual while not at work will be subject to the separate limits applying to any other person.

219 Employers are responsible for ensuring that dose limits are not exceeded for their employees, including trainees. Employers working with radiation should make sure that dose limits are not exceeded for people other than their own employees. However, this duty only relates to exposures resulting from their own work (see regulation 4(1)). Doses already received by these people as a result of work with ionising radiation on other sites must be taken into account.

220 Exposures received as a result of natural background radiation at normal levels are not considered in determining compliance with dose limits. Exposures resulting from work with any radioactive substance containing naturally occurring radioactive materials do have to be considered.

Assessing dose to the skin

221 Doses to organs and tissues are normally averaged over their volume for estimating effective dose and equivalent dose. The main exception is the skin, where the dose equivalent to the skin from external radiation, contamination on the skin or clothing and any radionuclides which may have accidentally entered wounds should be averaged over an area no greater than 1 cm^2 for comparison with the dose limit. For external radiation this will require careful consideration of the appropriate type and location of any dosemeter worn by an employee for assessing compliance with this limit. The aim should be to check that no area of the skin receives a dose exceeding the dose limit.

Dose limit for members of the public – exposures arising from patients undergoing medical diagnosis or treatment

222 People who may be significantly exposed to ionising radiation from being in close proximity to a patient receiving a medical exposure but who cannot be treated as carers and comforters are subject to the dose limits in Schedule 3 for 'any other person'. Such people may include friends who are unaware of the exposure they receive and children. Children are unable to give their consent and other relatives or friends will not have the opportunity to do so unless they visit the hospital. In general, doses received by relatives and friends in this situation are likely to be small but difficulties could arise in particular situations without some flexibility in the application of the public dose limit of 1 mSv a year effective dose.

223 In such cases, a hospital or clinic may take into account the provision in paragraph 6 of Schedule 3 for averaging the limit on effective dose for members of the public (5 mSv over five years). This provision should enable the hospital or clinic to make sensible arrangements for the release of patients who have been administered with a radiopharmaceutical and for restricting the exposure of relatives and friends. The British Institute of Radiology has published advice[17] on the timing of the release of such patients from hospital and the use of appropriate precautions. This advice relates to single administrations of radiopharmaceuticals and takes into account the restriction of consequential doses to members of the public.

Guidance   **12(1)**

Dose limit for members of the public – other exposures

224 The specific authorisation requirements under the Environmental Permitting (England and Wales) Regulations[18] and the Radioactive Substances Act[19] in Scotland are enforced by the relevant Environment Agency in England and Scotland and Natural Resources Wales. They will generally ensure that doses to members of the public arising from authorised discharge and disposal of radioactive material satisfy the public dose limit. However, this does not cover all practices carried out. The responsible employer will need to ensure that any additional exposure from direct external radiation does not cause the dose limit to be exceeded for the relevant representative person.

Overexposures

225 The term 'overexposure' is defined in regulation 2(1). Regulation 26 applies when an employer is informed or suspects that a person is likely to have received an overexposure. Regulation 27 provides an additional restriction for the continued exposure of any employee who has received an overexposure. This restriction must be treated as if it were a new dose limit for the remainder of the calendar year or five-year period for those subject to a dose limit under regulation 12(2).

Regulation   **12(2)**

(2)   Where an employer is able to demonstrate to the appropriate authority that, in respect of an employee, the dose limit specified in paragraph 1 of Part I of Schedule 3 is impracticable having regard to the nature of the work undertaken by that employee, the appropriate authority may in respect of that employee authorise the employer to apply the dose limits set out in paragraphs 8 or 9 of Schedule 3 and in such case the provisions of Part 2 of that Schedule will have effect.

Guidance   **12(2)**

Dose limitation for employees in special cases

226 In some cases, because of the special nature of the work carried out by an employee, it may not be practicable to comply with the limit of 20 mSv a year for adult employees. This situation may arise where there are skilled tasks which need to be undertaken by key specialist staff, including foreign nationals. Where the employer can demonstrate that this is the case, the provisions in Part 2 of Schedule 3 will apply.

227 Any employer wishing to apply the special dose limit of 100 mSv in five years (and no more than 50 mSv in any single year) for an individual employee under regulation 12(2) and Schedule 3 Part 2 would have to establish a need. Employers should consider the work that individual is likely to do in the year ahead and make a judgement as to whether or not that employee's annual dose would exceed 20 mSv, based on predicted doses. The judgement should take account of past experience and any plans for restricting future exposures from particular jobs as far as is reasonably practicable. The employer would also need to be satisfied that the employee's dose would not exceed the five-year limit or any other relevant dose limit (see paragraph 9 Schedule 3).

228 The application of this five-year dose limit to any particular employee is subject to a number of preconditions, which are set out in Schedule 3, such as:

(a)   consultation with the RPA and with the affected employees;

(b)   provision of information to the affected employees and the ADS, and seeking approval from HSE. Employers will also need to consult any appointed safety representatives;

(c)   investigation of any suspected exposures exceeding 20 mSv in a calendar year and notification to HSE;

(d)   a duty to review whether the five-year limit is still appropriate at least once every five years;

(e)   restrictions on reverting to an annual basis for the dose limit for that employee;

(f)   employers must record the reasons for choosing a five-year dose limit.

229   The appropriate authority can refuse such an application, for example requiring the employer to apply an annual limit of 20 mSv to the individual(s) concerned. However, this action is subject to appeal to the Secretary of State.

230   The purpose of the investigation required by paragraph 12 of Schedule 3 is mainly to check that any future exposure arising from the work is unlikely to result in an overexposure, ie that the five-year dose limit will still be met. If the investigation shows that this is not the case, the employer should consider what action is necessary to change the individual's working conditions. Note that if the individual has received an overexposure, an investigation under regulation 26 will be needed. There is no need for two investigations if an employee is made subject to the special five-year limit. The provisions of regulation 27 for any overexposure will apply to both the single-year and five-year dose limitation periods.

(3)   The steps taken by a relevant employer to comply with paragraph (1) in respect of members of the public must include an estimation of doses to members of the public from the relevant practice or practices carried out by the relevant employer in accordance with requirements regarding the estimation of doses as approved by the Executive from time to time.

(4)   In this regulation –

"appropriate authority" means

> *(a)   in relation to any activity carried out exclusively or primarily on nuclear premises, the ONR;*
>
> *(b)   otherwise, the Executive;*

"relevant employer" means an employer who is carrying out, or who intends to carry out, a relevant practice;

"relevant practice" means a practice to which regulation 6 or 7 applies.

Public dose estimation

231   Employers must make sure that the exposures members of the public receive as a result of the employer's work with radiation are as low as reasonably practicable (ALARP) and below the dose limit for other persons. Employers must also make realistic estimates of the exposure to the representative person for comparison with the dose limit in Schedule 3.

232   To estimate the exposure to external radiation, employers must estimate or measure the highest dose rates to which members of the public are exposed, or likely to be exposed. They must also identify the group of people who are most likely to receive the highest exposures. Such groups could be:

(a)   family members or friends who make regular visits to long-stay patients in a hospital where work with radiation is carried out;

(b)   family or friends who accompany a patient to a radiotherapy suite and sit in a waiting room;

(c)   dog walkers who exercise their dogs on public footpaths close to an industrial or medical radiography enclosure.

233 The overall principle in estimating the effective dose, E_{ext}, from external radiation in microSieverts per year ($\mu Sv\ y^{-1}$) to the representative person is to use:

(a) the highest dose rate to the representative person measured per hour where the dose rate is given by D ($\mu Sv\ h^{-1}$);

(b) multiplied by the number of hours (N) in one year over which the representative person is exposed.

234 Different groups of people will have different combinations of the above quantities. For example, a family member who makes regular visits to a long-stay patient may be exposed for more hours than a person who accompanies a patient for radiotherapy treatment, but is likely to be exposed to a lower dose rate. The exposure to the representative person should be estimated by the equation given in paragraph 236 for all the identified groups of people who are likely to receive the highest exposures. The highest result should be chosen as the dose to the representative person.

235 In the example described above of a public footpath adjacent to a radiography enclosure, where the most highly exposed person has been identified as a dog walker who walks their dog along the path every day for 10 minutes, the annual effective dose, E_{ext}, in μSv to the representative person is given by the equation:

$$E_{ext} = D \times 10/60 \times 365$$

where D = the highest measured dose rate at the boundary of the footpath ($\mu Sv\ h^{-1}$)

236 The effective dose from external radiation (E_{ext}) to the representative person is given by the equation:

$$E_{ext} = D \times N$$

Regulation 13 Contingency plans

(1) Where an assessment made in accordance with regulation 8 shows that a radiation accident is reasonably foreseeable (having regard to the steps taken by the employer under paragraph (3) of that regulation), the employer must prepare a contingency plan designed to secure, so far as is reasonably practicable, the restriction of exposure to ionising radiation and the health and safety of persons who may be affected by such accident.

Nature and scope of the plan

237 The aim of the contingency plan is to restrict exposures that arise from an accident so far as is reasonably practicable. The plan includes exposures both to the employees themselves and to others (including members of the public and emergency services workers) during the incident, clean-up and any associated recovery. Regulation 8 requires a radiation risk assessment to be carried out, which must include setting out the measures that need to be taken to restrict exposure in the event of an identifiable radiation accident.

238 The radiation risk assessment will identify all reasonably foreseeable accidents. This means accidents that are less than likely, but realistically possible. However, the action required under regulation 13(1) should be proportionate to the risk and likely magnitude of exposure. For example, small contained spillages of radioactive material and other incidents that could not result in exposures of concern for this task are not radiation accidents. An exposure of concern is where the accident, or actions such as clean-up, resulted or could result in a significant

exposure, ie an exposure which significantly exceeds normal planned exposures. It is therefore not proportionate to have a contingency plan in these circumstances and such incidents can be dealt with by having a spillage procedure or other protocol under usual risk control methods.

239 For radiation accidents where there is a risk of significant exposure, such as those involving industrial radiography or medical radiotherapy sources, a contingency plan must be prepared. The detail, scope and size of the plan may differ depending on the potential effect of the accident and should be proportionate to the potential impact. Planning for such effects means that thought has been given to:

(a)   the correct courses of action to take in an accident situation;
(b)   recording those actions in a plan;
(c)   training those identified to implement the plan;
(d)   providing appropriate equipment for dealing with the events or accidents and making sure this is available and maintained.

The RPA should be consulted about the plan.

240 An employer must have reasonable grounds for believing an event could occur in order to plan the contingency measures needed. Examples of events that could be reasonably foreseeable and could cause significant exposure and therefore require contingency plans are:

(a)   failure of specific control equipment such as shielding, electrical interlocks and warning devices;
(b)   failure of control measures such as fume cupboards, faulty PPE or misuse leading to exposure;
(c)   contamination spread from the controlled or supervised area;
(d)   fire leading to possible dispersal of radioactive material, or melting of lead shielding leading to substantial increases in dose rates;
(e)   faulty equipment, eg X-ray sets;
(f)   user error, for example correct procedures not being followed;
(g)   loss or theft of radioactive material.

241 In some accident scenarios (eg fires), it will be necessary for the plan to cover prior consultation with the emergency services. This is to make sure that all likely responders are aware of the role they will be expected to play, are prepared and able to control the radiation exposure of their own employees.

242 Certain dutyholders may have to plan for accident scenarios which could cause off-site emergencies. These may require specific plans and co-operation with local authorities. These duties are captured under the Radiation (Emergency Preparedness and Public Information) Regulations 2001.

Content of the plan

243 The level of detail in the plan should reflect the circumstances anticipated. Some plans may be generic where the same operations are conducted in different places at different times, but where the exposure conditions arising from a potential accident are likely to be similar. Site radiography and transport are two obvious examples.

Guidance   13(1)

244 In particular, the plan should identify:

(a)   who is responsible for putting the plan into effect;

(b)   what immediate actions for assessing the seriousness of the situation are necessary, for example the use of suitable radiation and contamination monitors;

(c)   what immediate mitigating actions are needed, for instance in clearing the accident area and establishing temporary means of preventing access to that area;

(d)   what emergency equipment is required to deal with identified accidents and where this can be found;

(e)   other sources of information and guidance, such as equipment manufacturers and contact details;

(f)   what PPE is needed and where it can be found;

(g)   what personal dosimetry requirements there are for people involved in controlling the accident;

(h)   what training is required for employees;

(i)   how to obtain radiation protection expertise so that proper judgements can be made about the seriousness of the situation and the measures necessary to recover from it;

(j)   under what circumstances to contact the emergency services and who is responsible for doing this;

(k)   what dosimetry follow-up is needed so that the people affected by the accident are identified and provision is made for their dose assessment.

Regulation   13(2)

(2)   An employer must ensure that –

(a)   where local rules are required for the purposes of regulation 18, a copy of the contingency plan made in pursuance of paragraph (1) is identified in those rules and incorporated into them by way of summary or reference;

(b)   any employee under the employer's control who may be involved with or affected by arrangements in the plan has been given suitable and sufficient instructions and where appropriate issued with suitable dose meters or other devices;

(c)   where appropriate, rehearsals of the arrangements in the plan are carried out at suitable intervals; and

(d)   if circumstances arise where it is necessary for some or all of the arrangements in the plan to be carried out –
   (i)   the cause of those circumstances is analysed to determine, so far as is reasonably practicable, the measures, if any, required to prevent a recurrence of such circumstances;
   (ii)   a record of such analysis is made and kept for at least 2 years from the date on which it was made; and,
   (iii)   any exposure which occurs due to the above circumstances is noted on any relevant dose record.

Guidance   13(2)

245 Employers must choose a suitable dosemeter or other device, taking account of the types of radiation and the dose rates likely to be encountered. While a dosemeter provided for routine dose assessment will often be suitable, other types may be more appropriate in some circumstances – for instance allowing immediate reading or giving high-dose-rate warning. Regulation 24 sets out requirements for dosimetry in cases of accidents and other incidents.

Guidance   **13(2)**

246 Where appropriate, the plan must be rehearsed. When assessing the need to rehearse the plan, consideration should be given to the:

(a) potential severity of the accident;
(b) likely doses that could be received by employees or others;
(c) complexity of the plan;
(d) number of people likely to be involved in its implementation;
(e) involvement of the emergency services;
(f) need to train employees and outside workers on what to do if an accident occurs;
(g) required frequency of plan rehearsal;
(h) implementation of lessons learnt from testing the plan.

247 If the contingency plan is used, the employer responsible for the plan should analyse and record this event. Where appropriate, the management and the RPA should work together on this analysis along with affected employees, their representatives, and if necessary, outside workers and their employers. The level of analysis should be proportionate to the incident that has taken place; this is not an investigation as in regulation 26. The analysis should consider:

(a) why the plan was used;
(b) what was the specific cause, or causes of the accident;
(c) whether any precautionary measures failed;
(d) if risk assessments need review;
(e) whether any general management failures were the cause, or contributed to the cause of the accident;
(f) who was involved and any injuries or exposures that occurred;
(g) the effectiveness of the contingency plan and whether it needs to be revised;
(h) what can be done to make sure lessons are learnt, results are shared and an event like this does not happen again.

248 Employers should encourage the recording and the proportionate analysis of near misses and sharing of the findings to help:

(a) identify trends;
(b) introduce corrective actions;
(c) prevent further accidents.

PART 3 Arrangements for the management of radiation protection

Regulation 14 Radiation protection adviser

Regulation 14(1)–(3)

(1)   Subject to paragraph (3), every employer engaged in work with ionising radiation must consult such suitable radiation protection advisers as are necessary for the purpose of advising the employer on the observance of these Regulations and must, in any event, consult one or more suitable radiation protection advisers with regard to the matters set out in Schedule 4.

(2)   Where an employer consults a radiation protection adviser pursuant to the requirements of paragraph (1) (other than in respect of the observance of that paragraph), the employer must appoint that radiation protection adviser in writing and must include in that appointment the scope of the advice which the radiation protection adviser is required to give.

(3)   Nothing in paragraph (1) requires an employer to consult a radiation protection adviser where the only work with ionising radiation undertaken by that employer is work specified in Schedule 1.

ACOP   14

249 To be suitable, an RPA will need to have the specific knowledge, experience and competence required for giving advice on the particular working conditions or circumstances for which the employer is making the appointment. In addition to the specific matters set out in Schedule 4, employers are required to consult an RPA where advice is necessary for compliance with the Regulations. This should include:

(a)   the radiation risk assessment required by regulation 8;
(b)   the designation of controlled and supervised areas as required by regulation 17, except where there is good reason to consider that such areas are not required, for example based on advice from the supplier of the radiation source or written guidance from an authoritative body;
(c)   the handling of the various investigations required by the Regulations;
(d)   the drawing up of contingency plans required by regulation 13;
(e)   the dose assessment and recording required by regulation 22.

Guidance   14(1)–(3)

General advice on RPAs

250 Employers must select suitable RPAs, one or more who have the required knowledge and experience for the employer's type of work. Possession of proof of competence does not necessarily make the holder suitable for the type of work under consideration.

Choosing a suitable RPA

251 An RPA should be suitable if they:

(a)   conform to the definition of an RPA as set out in regulation 2;

Guidance 14(1)–(3)

(b) have the required knowledge and experience relevant to the employer's type of work. This decision can be based on the RPA's working history.

252 The RPA cannot be expected to have all the specialist knowledge required by the employer. It is important that the RPA has the ability to recognise the limitations of their knowledge and experience in certain areas and have access to that specific information from other sources. For instance, having an understanding of the types of instrument to use in particular work situations does not mean that the RPA is an expert in instrumentation. Similarly, an RPA dealing with radon problems may well need access to expertise in ventilation to advise on adequate exposure restriction and to expertise in building mitigation for most indoor workplaces.

Consulting an RPA

253 Where employers need advice on compliance with the Regulations, this must be sought through consultation with an RPA. Unless Schedule 1 exempts a practice, an RPA must be consulted on the matters listed in Schedule 4 and also if the dose limitation scheme in Part 2 of Schedule 3 is to be invoked. Employers should obtain written confirmation of any key advice from the RPA as this will provide evidence of consultation.

254 The advice of the RPA should cover, where relevant, but not be limited to, the following:

(a) optimisation and establishment of appropriate dose constraints;
(b) plans for new installations and the acceptance into service of new or modified radiation sources in relation to any engineering controls, design features, safety features and warning devices relevant to radiation protection;
(c) categorisation of controlled and supervised areas;
(d) classification of workers;
(e) outside workers;
(f) PPE;
(g) workplace and individual monitoring programmes and related personal dosimetry;
(h) appropriate radiation monitoring instrumentation;
(i) quality assurance;
(j) arrangements for prevention of accidents and incidents;
(k) training and retraining programmes for exposed workers;
(l) investigation and analysis of accidents and incidents and appropriate remedial actions;
(m) employment conditions for pregnant and breastfeeding workers;
(n) preparation of appropriate documentation such as prior risk assessments and written procedures.

255 Involving the RPA in the radiation risk assessment (required by regulation 8, or in reviewing an assessment, required by regulation 3 of the Management Regulations) could be very useful as it will identify at the outset where the risks arise and what precautions and procedures are going to be necessary.

Appointment of RPAs

256 An employer, who intends to consult a suitable RPA for the purposes of these Regulations, must appoint the RPA in writing. This is intended to formalise the terms of the arrangement between the RPA and the employer, by specifying the scope of the advice the RPA is required to give.

Guidance 14(1)–(3)

257 An appointment in writing is not required for an initial consultation, for example where the RPA is only asked to advise the employer if, given the nature of the work with ionising radiation, formal consultation is necessary.

258 RPAs can be part-time or full-time employees or consultants, depending on the extent of demand for continuing advice. For smaller organisations, such as those whose work with ionising radiation is confined to using an xrf analyser, a consultant may well suffice. In this case, detailed advice required at the early stages of setting up could be followed by an access arrangement on an as-and-when basis. A large or complex organisation may need a number of RPAs or an RPA body (one which meets HSE's Criteria of Core Competence)[20] if the range of tasks or scope and complexity of the work with ionising radiation is such that one person could not reasonably provide the breadth of advice or the functions required. These RPAs might have different individual areas of expertise for different purposes.

Availability of RPA advice

259 RPAs appointed on a continuing basis should be available for consultation whenever required under the access arrangement agreed with the employer. They do not need to be present every time that work with ionising radiation takes place.

Exemption from the need to appoint an RPA

260 Employers whose work with ionising radiation comes within any of the descriptions in Schedule 1 are not required to consult and do not need to appoint an RPA. However, they may wish to consult an RPA, at least initially, for checking or reassurance purposes.

Regulation 14(4)

(4) The employer must provide any radiation protection adviser appointed by it with adequate information and facilities for the performance of the radiation protection adviser's functions arising from their consultation or appointment under this regulation.

Guidance 14(4)

Information and facilities for the RPA

261 Employers must make sure that their RPAs have access to all the information and facilities that they need to perform their duties effectively. The information should include a clear statement of the scope of the advice each RPA is required to give. The facilities may need to include appropriate support services, eg secretarial, unless the RPAs provide their own. Where there is a potential for emergencies with off-site consequences, it is advised that specialised radiation protection units should normally be provided to support the RPA(s).

Regulation 15 Information, instruction and training

Regulation 15(1)(a)

(1) Every employer must ensure that –

(a) those of its employees who are engaged in work with ionising radiation are given appropriate training in the field of radiation protection and receive such information and instruction as is suitable and sufficient for them to know –
(i) the risks to health created by exposure to ionising radiation as a result of their work;
(ii) the general and specific radiation protection procedures and precautions which should be taken in connection with the work with ionising radiation to which they may be assigned; and

(iii)   the importance of complying with the medical, technical and administrative requirements of these Regulations;

(b)   adequate information is given to other persons who are directly concerned with the work with ionising radiation carried on by the employer to ensure their health and safety so far as is reasonably practicable;

Guidance   15(1)(a)–(b)

262 All those involved in work or affected by work with ionising radiation, including management and outside workers, need to know how to work safely and reduce risk to their health. They must be trained to help develop and sustain a commitment to restricting exposure wherever it is reasonably practicable.

263 Training must be appropriate to the nature of the work and designed to meet the specific needs of employees. Employers must provide clear instruction and training on both the general operational and working conditions of the practice and the particular setup to which the employee is assigned.

264 Employers must provide training to make sure employees are competent where a system of work or PPE is provided to restrict exposure (as required by regulation 9). Training is also needed where the employer arranges for employees to perform particular functions required by these Regulations, for example to make entries in radiation passbooks for outside workers (regulation 19).

265 Some employees may not be closely involved with the work but must be given suitable and proportionate information or instruction to avoid being unnecessarily exposed to ionising radiation. The duty under this regulation complements the general duties on information, capabilities and training in regulations 10 and 13 of the Management Regulations.

266 Employees involved in work with ionising radiation need to understand the results of the general risk assessment and what this means for them. They must be made aware of the main risks, including the risk of accidental exposures, and the control measures they must follow to prevent or reduce those risks.

267 Employees must also be instructed on their responsibilities. They must co-operate with their employer and follow the health and safety training provided.

268 The duty in regulation 15 is placed on all employers in relation to their employees; it is not limited to employers who are undertaking the work with ionising radiation. If a contractor carries out work, other than work with ionising radiation, on the site of an employer who is undertaking work with ionising radiation, both employers will be expected to share information under regulation 16. Effective co-operation will ensure that the contractor can inform their own employees about any radiation risks associated with their work on the site and any preventative measures they must take to avoid those risks.

What is required?

269 Training must be effective and employers should check its adequacy. Checks could include:

(a)   requesting feedback from employees and supervisors;
(b)   making sure that employees are working as they have been trained;
(c)   monitoring the effect of training on accidents or near misses.

270 Employers should consider whether the training delivered its aims and objectives and, if it hasn't, make changes to the programme. Reviewing training material regularly will make sure that it remains current. Employers must consider

Guidance 15(1)(a)–(b)

remedial training if lack of competence is identified as the cause of an incident or has contributed to it. Keeping training records will help to identify when refresher training is needed.

271 Individuals should know when they need to seek help and where they should find it. Employers are advised to consult their RPAs when planning their information, instruction and training needs.

Regulation 15(1)(c)

(c) *its female employees who are engaged in work with ionising radiation are informed of the possible risk arising from ionising radiation to the foetus and to a nursing infant and of the importance of their informing their employer in writing as soon as possible –*

(i) *after becoming aware of their pregnancy; or*
(ii) *if they intend to breast feed an infant;*

Guidance 15(1)(c)

Information for employees who may become pregnant or start breastfeeding

272 Employees should notify their employer in writing as soon as they become pregnant or if they are breastfeeding so the employer can put in place any special protection required under regulation 9(6), unless working conditions are such that no special protection would be necessary.

273 As part of an employer's general duty to assess risks under the Management Regulations, they must consider the preventative and protective measures they will implement to reduce, remove or control risks to pregnant or breastfeeding workers. The risks and control measures must be communicated to female workers.

274 Information on risks and control measures from exposure to ionising radiation, particularly relating to the possible risk to the foetus, could include any relevant guidance from HSE, eg the leaflets *Working safely with ionising radiation: Guidelines for expectant or breastfeeding mothers* (INDG334) and *New and expectant mothers who work: A brief guide to your health and safety* (INDG373)[14] or the relevant doctor.

Regulation 15(1)(d)–(e)

(d) *any employees engaged in work in a controlled area (as designated under regulation 17) are given specific training in connection with the characteristics of the workplace and the activities within it;*
(e) *the giving of training and information under this regulation is repeated at appropriate intervals and documented by the employer.*

Guidance 15(1)(e)

Repeating of training at regular intervals

275 Refresher training should be scheduled at regular intervals to maintain competence levels. In addition, employers should review employees' capabilities and provide additional or refresher training for employees as needed.

276 Employers should determine the frequency of refresher training. It may be possible for such training to be given as part of other health and safety updates. A realistic risk-based approach to refresher training is needed to make sure that an employee's knowledge and awareness are maintained.

277 If new equipment is brought in or working practices change, staff will require further training.

<table>
<tr><td>

Regulation 15(2)

</td><td>

(2) In addition to the requirements in paragraph (1), every employer who is engaged in work with ionising radiation involving a high-activity sealed source must ensure that the information and training given to employees involved in such work includes –

(a) specific requirements for the safe management and control of high-activity sealed sources for the purpose of preparing such employees for any events which may affect their radiation protection;

(b) particular emphasis on the necessary safety requirements in connection with high-activity sealed sources; and

(c) specific information on the possible consequences of the loss of adequate control of high-activity sealed sources.

</td></tr>
</table>

Guidance 15(2)

278 Training for high-activity sealed sources (HASS) should include the following:

(a) training and instruction in the precautionary measures required when dealing with HASS;

(b) training and instruction for those carrying out maintenance checks, including leak tests as designated under regulation 28(3);

(c) training and instruction in fire prevention with regard to sources;

(d) training in the appropriate management procedures for handling any lost, loose or detached sources discovered;

(e) training and instruction in the emergency procedures which must be followed in the event of any foreseeable accident that could result in damage to the sources, for example fire;

(f) information on the risks and potential effects on people, including those who may come too close to HASS, touch them or pick them up, particularly if the sources are damaged;

(g) information on the possible serious consequences of the loss of adequate control of HASS;

(h) training and instruction in the emergency procedures which must be followed in the event of loss, theft or unauthorised use of HASS;

279 Employers may find it helpful to include past accident or incident scenarios involving the loss or theft of HASS in the training as a way of highlighting the risks.

Reviewing training of employees working with high-activity sealed sources

280 Where the training of employees working with HASS was carried out before these Regulations came into force, that training must be reviewed to make sure it remains sufficient and suitable. More specific legal duties on employers have been introduced by these Regulations and therefore employers may find that the original training has to be revised.

Regulation 16 Co-operation between employers

Regulation 16

Where work with ionising radiation undertaken by one employer is likely to give rise to the exposure to ionising radiation of the employee of another employer, the employers concerned must co-operate by the exchange of information or otherwise to the extent necessary to ensure that each such employer –

(a) has access to information on the possible exposure of their employees to ionising radiation; and

(b) is enabled to comply with the requirements of these Regulations in so far as their ability to comply depends upon such co-operation.

Guidance 16

Employers sharing the same workplace

281 If two employers share the same workplace, whether on a temporary or permanent basis, they have legal duties to co-operate with each other. Everyone must understand their role in ensuring health and safety regulations are observed. When information has been shared, employers should check that it has been understood and that agreed procedures are followed.

282 The aim of co-operation should be to co-ordinate the measures taken to comply with statutory duties and to inform each other of the risks to employees arising out of their work (site employers also need to pass on clear information about general risks under regulation 12 of the Management Regulations). This includes all parties exchanging information on:

(a) health and safety risks;
(b) measures taken to control the risks and safe systems of work;
(c) fire precautions, including information to enable outside workers to identify the person nominated to implement evacuation procedures within premises where the work is taking place.

283 Compliance with regulation 12 of the Management Regulations should be sufficient to satisfy the requirements of regulation 16 of IRR17. Information on health and safety risks that should be shared includes:

(a) details relating to controlled areas that could be entered by employees of either employer;
(b) relevant contingency arrangements for action to prevent or mitigate the consequences of any radiation accident (see regulations 8(3), 13 and 19(2)).

284 As regards work in buildings with radon atmospheres subject to these Regulations, the employer controlling the premises should notify other employers working in the building.

Sharing dose information

285 Where an employee has several employers, those employers must co-operate to make sure dose limits are applied to the total dose received by the worker. For example, in the health sector, there are several groups of workers (such as cardiologists) who may have multiple employers (eg an NHS trust and one or more private hospitals) or who may act in a self-employed capacity for some of the year.

286 If a worker exceeds the dose limit, the employer at the particular point when the dose limit is exceeded is deemed to be responsible. Every employer must be aware of exposures received elsewhere to make sure dose limits are complied with. For example, a worker may have received a whole body dose of 19.5 mSv while working for a public sector employer. They then carry out work in the private sector, where they receive a dose of 1 mSv, taking the total dose above the current dose limit. The private sector employer would be in breach of regulation 12.

287 It is important for employers to share information on the total dose received otherwise they will have difficulties in making the correct decisions on the need for classification and ensuring compliance with relevant dose limits.

288 If an employee has several employers or is self-employed on a part-time basis, employers must have a method of finding out where the dose was received.

Outside workers

289 Outside workers may be at particular risk, as they may be unfamiliar with the undertaking's procedures, rules, hazards and risks.

290 Employers of outside workers must find out from the employer in control of the work area what risks and additional training needs are associated with the services the outside worker will perform, well before those services are due to start.

291 It is important to share this information when contracts and arrangements are still being discussed. This should help to make sure the right workers are chosen to carry out the services and that, before they start, those workers are given the relevant information and training so that they can protect themselves and others properly.

292 Employers in control of the work area also need information about the outside worker (eg dose history and training needs) before they start work on site (see regulation 19(3)(b)). In most cases, both employers should consult their RPAs on this matter.

293 Employers of outside workers will require the following information:

(a) the detail of the actual work to be carried out;
(b) the type of likely radiation exposure;
(c) an estimate of the total dose likely to arise from the work;
(d) the work procedures that will be required to keep doses as low as reasonably practicable (including any use of special protective equipment);
(e) the risks associated with the work and the precautions that should be taken;
(f) any local restrictions that will be applied;
(g) the local rules that apply in the other employer's site (including emergency arrangements and contingency plans);
(h) radiation protection supervisor (RPS) appointments;
(i) any relevant dose constraints and associated local investigation levels.

294 The amount of detail depends on the complexity and duration of the work.

295 Employers should know whether additional dose assessments will be necessary for their outside workers. For example, assessment of external radiation exposures to the eye or an extremity, or of internal exposure, may be necessary. If additional dose assessment is needed, the employers should agree between them who will make the necessary arrangements. The outside worker's employer should be aware of the methods of dose estimation that the other employer will use. In the event of an incident, additional dose estimates may be required (see regulation 24).

296 If the work will take place outside normal hours, the employers should exchange information on emergency contacts.

PART 4 Designated areas

Regulation 17 Designation of controlled or supervised areas

Regulation 17(1)

(1) Every employer must designate as a controlled area any area under its control which has been identified by an assessment made by that employer (whether pursuant to regulation 8 or otherwise) as an area in which –

(a) it is necessary for any person who enters or works in the area to follow special procedures designed to restrict significant exposure to ionising radiation in that area or prevent or limit the probability and magnitude of radiation accidents or their effects; or

(b) any person working in the area is likely to receive an effective dose greater than 6 mSv a year or an equivalent dose greater than 15 mSv a year for the lens of the eye or greater than 150 mSv a year for the skin or the extremities.

ACOP 17(1)

297 Special procedures should always be necessary to restrict the possibility of significant exposure. Employers should designate controlled areas in cases where:

(a) the external dose rate in the area exceeds 7.5 μSv per hour when averaged over the working day;

(b) the hands of an employee can enter an area and the 8-hour time average dose rate in that area exceeds 75 μSv per hour;

(c) there is a risk of spreading significant radioactive contamination outside the working area;

(d) it is necessary to prevent, or closely supervise, access to the area by employees who are unconnected with the work with ionising radiation while that work is under way;

(e) employees are liable to work in the area for a period sufficient to receive an effective dose in excess of 6 mSv a year.

In addition, an area should be designated as a controlled area if the dose rate (averaged over a minute) exceeds 7.5 μSv per hour and employees untrained in radiation protection are likely to enter that area, unless the only work with ionising radiation involves a radioactive substance dispersed in a human body and none of the conditions in (a) to (e) in paragraph 307 apply.

Guidance 17(1)

298 The designation of a controlled area is also likely to be required when the instantaneous dose rate exceeds 100 μSv per hour even though the dose rate, when averaged over a working day, is less than 7.5 μSv per hour. In addition, an area must be designated as a controlled area if the dose rate (averaged over a minute) exceeds 7.5 μSv per hour and the work being carried out is site industrial radiography.

Responsibility for designating an area

299 The employer in control of an area is responsible for designating that area. Usually this is the employer who is in overall control of the site, eg the main site employer who carries out work with ionising radiation. Where that employer assigns temporary control of the area to a contractor, that contractor is responsible for deciding whether or not to designate the area as a controlled area. Contractors must co-operate with site employers, either to inform them about the extent of any controlled area they create, or to pass on information about the risks arising from their work with ionising radiation.

300 Paragraph 249 advises that the employer must normally consult an RPA about the need to designate a controlled or supervised area.

Purpose of designating controlled areas

301 The main purpose of designating controlled areas is to make sure that the measures provided under regulation 9(1) are effective in preventing or restricting routine and potential exposures. This is achieved by controlling who can enter or work in such areas and under what conditions. Employers must designate controlled areas where people entering the area need to follow special procedures. Such procedures could take the form of a detailed system of work setting out how the tasks should be carried out in a way that restricts significant exposure.

Designation as a controlled area on the basis of special procedures

302 The employer's risk assessment will indicate where special procedures are necessary to restrict exposure in addition to the physical control measures required by regulation 9. Employers should designate an area as a controlled area if these procedures are specific to an area and require particular instructions to be followed by those who enter or work in the area (or untrained people to be excluded unless under close supervision).

303 The ACOP guidance in paragraph 297 advises on situations where a controlled area must be designated.

304 Employers must designate an area as a controlled area if any person is likely to receive an annual effective dose in excess of 6 mSv, or an annual equivalent dose greater than one of the other values specified in regulation 17(1)(b) as a result of work in that area.

305 Employers must put in place special procedures to prevent accidental exposures when people enter high-dose-rate shielded enclosures or plant. An example is where it is necessary for employees to follow a defined procedure involving the use of a suitable dose rate meter to check that a radiation source is safe before entry.

306 If special procedures are necessary, employers must designate the area as a controlled area whether or not the dose rate is above 7.5 µSv per hour.

307 In deciding whether or not a controlled area is needed, employers should consider:

(a) which people are likely to need access to the area;
(b) the level of supervision required;
(c) the nature of the radiation sources in use and the extent of the work in the area;

(d)   the likely external dose rates to which anyone can be exposed;

(e)   the likely periods of exposure to external radiation;

(f)   the physical control methods already in place, such as permanent shielding and ventilated enclosures;

(g)   the importance of following a procedure closely in order to avoid receiving significant exposure;

(h)   the likelihood of contamination arising and being spread unless strict procedures are closely followed;

(i)   the need to wear PPE in that area;

(j)   maximum doses estimated for work in the area.

308   In addition to the circumstances described in paragraphs 306 to 307, employers must designate an area as controlled if:

(a)   access is foreseeable to that area by people, such as office staff, whose work does not normally involve ionising radiation;

(b)   normal control measures for an area have to be suspended for work such as maintenance or source changing;

(c)   people are likely to be exposed to significant levels of surface or airborne contamination in the area, in excess of appropriate derived working levels or derived air concentrations;

(d)   RPE must be worn while working in the area.

Circumstances where designation is unlikely to be needed

309   Employers in control do not need to designate an area as a controlled area simply because people work under a general system that reflects good practice in that sector. This kind of system of work is not regarded as a 'special procedure', eg a controlled area would not normally be required where:

(a)   work is routine and special precautions are not required, for example work in the vicinity of a fixed radiation gauge (except maintenance work); or

(b)   work is carried out with low levels of radionuclides of low radiotoxicity inside efficiently ventilated enclosures (for example fume cupboards) or on a laboratory bench and only routine precautions are expected, such as the use of lined trays to contain spillage and the use of disposable protective gloves.

310   Places which cannot physically be entered do not need to be designated. This includes areas where it is not reasonably foreseeable that a person, or part of a person, can enter or be present in that area.

311   If the only person in that area who is exposed to ionising radiation is a person undergoing medical examination or treatment (see regulation 3(3)), then the area does not need to be designated. However, employers must consider the possibility of exposures, including accidental exposures, of other members of the public and members of staff.

Designation in the case of radionuclides in the human body

312   Designation will probably be necessary in a few limited cases, for example where a patient remains in a hospital or clinic after the therapeutic administration of a radiopharmaceutical and:

(a)   the work with ionising radiation involves a radioactive substance dispersed in a human body where that substance emits gamma rays and the product of activity and total gamma energy per disintegration exceeds 150 MBq.MeV; or

(b)   the patient is undergoing brachytherapy using a gamma source or a high-energy beta ray source.

Designation of an area taking account of physical features

313   When considering the extent of any controlled area, take account of the physical boundaries, such as walls and partitions around the working area. If it is more convenient to use these boundaries (for example because of the need to control access), they can be used rather than a smaller part of the area where dose rates or contamination levels are significant.

314   These boundaries should not be too far from the area of concern to allow proper control of the area. Once such an area has been designated, it is subject to all the legal requirements applying to controlled areas under regulations 18, 19 and 20. The whole of a room does not need to be designated as a controlled area provided that the necessary restrictions can be applied to that part of a room or laboratory where it is necessary to prevent or restrict access.

Temporary de-designation of controlled areas

315   If the periods during which work with ionising radiation takes place are clearly defined, follow a regular pattern, or are only intermittent, employers can de-designate on a regular basis. An example of de-designation would be to allow cleaners to have routine access where this is appropriate. Employers must take sufficient steps to remove the need for designation of the area by, for example, isolating an X-ray generator from the power supply or removing any radioactive substances or making them safe. Employers must summarise these steps in local rules prepared in compliance with regulation 18(1).

Temporary designation of an area as controlled for a particular task

316   An employer may decide it is unnecessary to designate an area as a controlled area because employees do not enter that area and physical safeguards prevent accidental exposures. If contractors have to enter the area for particular tasks, such as source changing or maintenance, it may be necessary to designate such areas temporarily under specified conditions.

Area designation during transport or movement on site

317   The dose rates inside and outside vehicles used to carry packages containing radioactive material may be in excess of the levels at which areas may require designation. For example, designated areas may be required in the cab of a vehicle or around a vehicle or package during a temporary stop.

Designation on the basis of annual dose

318   In practice, it is often difficult to predict annual doses received by employees from knowledge of dose rates in working areas because:

(a)   dose rates are seldom constant over long time periods and within the physical boundaries of areas;

(b)   there are significant variations in the pattern of work for individuals;

(c)   the duration of an individual's exposure in the areas may be difficult to estimate.

319 Consequently, the expected annual dose is not likely to be the main criterion in most cases for deciding whether an area needs to be designated as a controlled area. One exception might be areas in radon-affected workplaces where high radon levels are known to occur and no special procedures need to be followed by employees. Also, where employees work for about 2000 hours a year in an area where the external dose rate routinely exceeds 3 µSv per hour, that area may need to be designated as a controlled area because that individual would be likely to receive a dose greater than 6 mSv a year.

(2)   An employer must not intentionally create in any area conditions which would require that area to be designated as a controlled area unless that area is for the time being under the control of that employer.

Intention to create conditions

320 The area concerned may be under the control of another employer or outside the boundary of a site. In either case, the purpose of this provision is that employers do not plan or carry out work that gives rise to external dose rates or levels of contamination which require access to be restricted in an area they do not control. An example would be site radiography work adjacent to a footpath where people walking past the area could be significantly exposed to ionising radiation. If these conditions arise in an area as a result of an unforeseen event, the employer would not be considered to have intentionally created such conditions.

Designation of an area after an accident

321 An accident might create conditions which would warrant the designation of a controlled area in a place where the employer does not normally have control. In such cases, that employer should, where possible, have the access to that area restricted until the situation returns to normal or until the emergency services take over control of it. Employers should include these arrangements in the contingency plan under regulation 13.

(3)   An employer must designate as a supervised area any area under its control, not being an area designated as a controlled area –

(a)   where it is necessary to keep the conditions of the area under review to determine whether the area should be designated as a controlled area; or

(b)   in which any person is likely to receive an effective dose greater than 1 mSv a year or an equivalent dose greater than 5 mSv a year for the lens of the eye or greater than 50 mSv a year for the skin or the extremities.

Designation of supervised areas

322 The decision to designate an area as a supervised area depends both on the assessment of likely doses in that area and the probability that conditions might change. For example, where an area needs to be kept under review because of the possibility that radioactive contamination might spread, you must designate the area as a supervised area.

323 It is not necessary to designate a supervised area outside every controlled area. For example, if a controlled area has been designated on the basis of external dose rate, and conditions in adjacent areas are unlikely to alter significantly, a supervised area will not be necessary unless a person is likely to receive a dose in excess of 1 mSv a year in those adjacent areas.

324 In some laboratories only small quantities of unsealed radioactive substances are used. In these situations it may not be appropriate to designate the room as a

controlled area to make sure that specific procedures are followed by those who enter or work there. In such a laboratory, there will be general arrangements for preventing and cleaning up any contamination arising from spillages. Employers must designate at least part of the laboratory as a supervised area if contamination could build up over a period of some weeks as a result of not following these arrangements. The part of the laboratory designated as a supervised area should be chosen to reduce the risk of contamination, for example around a fume cupboard.

325 Employers can choose boundaries for the supervised area which are convenient. Once such an area has been designated it is subject to all the legal requirements applying to supervised areas.

Regulation 18 Local rules and radiation protection supervisors

Regulation 18(1)–(2)

(1) For the purposes of enabling work with ionising radiation to be carried on in accordance with the requirements of these Regulations, every employer engaged in work with ionising radiation must, in respect of any controlled area or, where appropriate having regard to the nature of the work carried out there, any supervised area, make and set down in writing such local rules as are appropriate to the radiation risk and the nature of the operations undertaken in that area.

(2) Local rules must identify the main working instructions intended to restrict any exposure in that controlled or supervised area.

Guidance   18(1)–(2)

Responsibility for local rules

326 The employer carrying out the work with ionising radiation is responsible for making sure local rules are prepared, as they have overall responsibility for providing control measures to restrict exposure under regulation 9.

327 Where more than one employer works in a controlled area, each employer has a duty to prepare local rules, even if these are adapted from the site occupier's local rules.

Local rules for supervised areas

328 Local rules should be in place for supervised areas where the employer has to instruct employees about general arrangements to prevent accidents, or to restrict exposure in that area. Examples include:

(a) maintenance and cleaning of an area where unsealed sources are used;
(b) arrangements for putting a contingency plan into effect in the event an accident.

Nature of local rules

329 The details given in these rules should be appropriate to the nature and degree of the risk of exposure to ionising radiations. The rules must cover work in normal circumstances and also the particular steps needed to control exposure in the event of a radiation accident, as set out in the contingency plan required by regulation 13.

330 Local rules for a controlled area must include a summary of the arrangements for restricting access into that area, including the written arrangements covering those who are not classified persons.

Guidance 18(1)–(2)

331 Local rules set out the arrangements for restricting exposure in a particular area, usually controlled areas, and where appropriate, supervised areas. They can vary considerably in detail and format, depending on the complexity of the work with ionising radiation. Local rules can include instructions, booklets or circulars, but should contain the information listed in paragraph 336.

332 The employer's general management arrangements for complying with IRR17 form part of the general health and safety arrangements required by regulation 5 of the Management Regulations. Those arrangements may set out management responsibilities for radiation protection and include arrangements for monitoring or auditing the measures established to comply with these Regulations.

333 There is no need to duplicate those arrangements in local rules. However, employers may wish to cross-refer to the main features relevant to a particular area, or even to repeat or attach a copy of those general arrangements to a centrally held copy of the local rules for this purpose.

334 Local rules might be held in an electronic system. However, readily available paper copies may be needed for the RPS and staff working in the area, especially where the work is carried out away from the employer's base location.

335 Employers must obtain advice from an RPA about the local rules and their content (one of the requirements for controlled and supervised areas covered by Schedule 4). They may also need to consult any established safety committee or appointed safety representative(s).

Local rules

Essential contents
336 Local rules should contain the following information:

(a) the dose investigation level specified for the purposes of regulation 9(8);
(b) identification or summary of any contingency arrangements indicating the reasonably foreseeable accidents to which they relate (regulation 13(2));
(c) name(s) of the appointed RPS(s) (regulation 18(5));
(d) the identification and description of the area covered, with details of its designation (regulation 19(1));
(e) a summary of the working instructions appropriate to the radiological risk associated with the source and operations involved, including the written arrangements relating to non-classified persons entering or working in controlled areas (regulation 19(3)).
(f) where an employer has detailed written working instructions contained within operations manuals or work protocols, it will usually be sufficient for the local rules to refer to the relevant sections of these documents. However, the employer must make sure the way in which these are summarised in local rules is adequate for the purposes of regulation 18.

Optional contents
337 Employers may also find it useful to include a brief summary or reference to the general arrangements in that area for:

(a) testing and maintenance of engineering controls and design features, safety features and warning devices;
(b) radiation and contamination monitoring;
(c) examination and testing of radiation monitoring equipment;
(d) personal dosimetry;
(e) arrangements for pregnant and breastfeeding staff;

Guidance   18(1)–(2)

(f) details of significant findings of the risk assessment, or where it can be found;

(g) a programme for reviewing whether doses are being kept as low as reasonably practicable and local rules remain effective;

(h) procedures for initiating investigations etc;

(i) procedures for contacting and consulting the appointed RPA;

(j) details of the management and supervision of the work;

(k) procedures for ensuring staff have received sufficient information, instruction and training.

338 Where these matters are not already included in general health and safety management arrangements, consider including the information on relevant aspects of the health and safety management system in an annex to the local rules. This will supplement the key points mentioned in the essential contents section. Requirements on the health and safety management system are also subject to regulation 5 of the Management Regulations.

Level of detail required

339 Local rules are intended to focus on arrangements that are needed in a particular area. The level of detail required in the written local rules will depend on the nature of the work in that area. Therefore, employers should consider the:

(a) risk arising from the exposure in normal operations;

(b) complexity of the work being done;

(c) likelihood of a failure or accident and the magnitude of any resulting exposures;

(d) level of information, instruction and training given to those working in the area and the extent of their knowledge and experience.

These matters would normally be considered as part of the employer's risk assessment.

Making local rules effective

340 Local rules are effective if they:

(a) are brief and concentrate on the activities which give the greatest risks;

(b) focus on the key working instructions to be followed by everyone to keep radiation doses as low as reasonably practicable;

(c) are 'local' by referring directly to work at a particular location and by taking account of the environment in and around the working area;

(d) contain clear instructions which reflect actual work practice rather than an ideal which is not attainable;

(e) are reviewed periodically to check that they remain relevant.

In preparing local rules, employers should balance providing the necessary details with the ability of employees and others affected to read, understand and take appropriate action. If the local rules are too long or unclear, they may not be read by those entering or working in the area and will not be effective.

Regulation   18(3)

(3) An employer must take all reasonable steps to ensure that any local rules which are relevant to the work being carried out are observed.

Guidance   18(3)

341 Regulation 5 of the Management Regulations requires employers to put a management and supervisory system in place to ensure the effectiveness of the health and safety measures. This includes the control measures necessary for compliance with IRR17, supplemented by other steps to ensure that local rules are

Guidance 18(3)

followed. Employers must make sure that employees, outside workers and others are following local rules.

342 Where a small team of employees spends long periods working away from the base location, it may be appropriate for a regional manager to arrange for periodic audits of the way the work is being carried out.

Regulation 18(4)

(4) An employer must ensure that any relevant local rules are brought to the attention of those employees and other persons who may be affected by them.

Guidance 18(4)

343 Local rules must be available to employees and other people involved or likely to be affected at or near the area concerned, eg by displaying sections relevant to particular operations in the area, or immediately adjacent to it. They may be seen as a collection of parts relating to different areas, so employees only need to see relevant parts relating to the areas in which they work.

344 It may be helpful for managers responsible for updating and reviewing local rules (and appointed safety representatives) to hold a complete set of local rules for reference at one central point.

Regulation 18(5)

(5) An employer must –

(a) appoint one or more suitable radiation protection supervisors for the purpose of securing compliance with these Regulations in respect of work carried out in any area made subject to local rules;

(b) set down in the local rules the names of such radiation protection supervisors; and

(c) provide the means necessary for the radiation protection supervisor to perform their role.

Guidance 18(5)

General advice on radiation protection supervisors

345 The appointment of a radiation protection supervisor (RPS) under this regulation supplements, but is not a substitute for, the general requirements in regulations 5 and 7 of the Management Regulations for monitoring and health and safety assistance. The RPS has a crucial role to play in helping to make sure the arrangements made by the employer are adhered to, and in particular supervising the arrangements set out in local rules. However, the legal responsibility remains with the employer and cannot be delegated to the RPS.

Suitability for appointment as an RPS

346 An employee who is appointed as an RPS should:

(a) know and understand the requirements of the Regulations and local rules relevant to the work with ionising radiation;

(b) command sufficient authority from the people doing the work to allow them to supervise the radiation protection aspects of that work;

(c) understand the necessary precautions to be taken and the extent to which they will restrict exposures;

(d) be given sufficient time and resources to carry out their functions;

(e) know what to do in an emergency.

347 RPSs should receive appropriate training under regulation 15 to carry out the task adequately. The training will need to reflect the complexity of the work being done. This should include periodic refresher training to maintain competency levels and if changes are made to local rules.

348 In general, an RPS will be an employee of the employer carrying out the work with ionising radiation. They will usually be in line management positions, closely involved with the work being done, to allow them to exercise sufficient supervisory authority. Additionally, the role may be carried out by a radiation protection unit established within the employer's business or by an RPA.

349 In some situations, for example where a contractor is carrying out work on the site of another employer, ie an outside worker, it may be appropriate to appoint one of that site operator's employees as the RPS, eg where the site operator is an employer who works with radiation regularly and the contractor rarely works with ionising radiation. However, as the RPS undertakes a supervisory role, both employers may need to make suitable contractual arrangements for such an appointment.

Number of RPSs required

350 It may not always be necessary for an RPS to be present all the time. To decide how many RPSs are required, employers will need to take account of the range and complexity of the work and the number of different locations to be covered. Cover should be sufficient for the work being done, taking account of factors such as peripatetic work away from the base location of the company, shift work, and absence due to sickness, training and holidays. The number of RPSs must be sufficient to make sure that the work is adequately supervised.

Appointment of the RPS

351 An RPS should be clear about the role they are expected to fulfil. This should be confirmed in writing so there is no confusion about the work expected of them. Employers may wish to display the name of the relevant RPS on notice boards at fixed locations where the work is carried out.

Consultation with the RPA

352 Employers should consider the need for advice from the RPA about the suitability of RPS appointments.

Regulation 19 Additional requirements for designated areas

(1)   Every employer who designates any area as a controlled or supervised area must ensure that any such designated area –

(a)   is adequately described in local rules; and
(b)   has suitable and sufficient signs displayed in suitable positions warning that the area has been so designated and indicating the nature of the radiation sources and the risks arising from such sources.

(2)   A controlled area must be physically demarcated or, where this is not reasonably practicable, delineated by some other suitable means.

Description of controlled and supervised areas

353 Employers who designate such areas are responsible for describing them in local rules. Contractors carrying out work with ionising radiation within areas designated by other employers are not required to include a description of the area in their local rules.

Guidance   19(1)–(2)

354 Where it is not appropriate to provide local rules for a supervised area (see regulation 18(1)), that area could be described either in the local rules of an adjacent controlled area or, if this is not possible, in the centrally held copy of the local rules for the site.

355 Designated areas are usually described by reference to fixed features, such as walls. If the radiation source is mobile, the area(s) may be described generically, for example by reference to distances from the source or, if necessary, to distances from other objects irradiated by the source.

Physical demarcation of controlled areas

356 The main purpose of physically demarcating a controlled area is to restrict unauthorised access. To make sure suitable means of restricting access have been provided, employers must consider the nature of the work and how likely the means provided will restrict access only to people who are authorised to enter. To be effective, the method of demarcation must clearly indicate the extent of the controlled area, so there is no doubt.

357 In most cases physical features can be used, for example existing walls and doors. Employers should provide temporary barriers where physical features cannot be used, and these should be supervised where reasonably practicable. In such cases, the areas must be clearly delineated by other suitable means so employees (and other people as necessary) are aware that these areas exist.

358 Examples of situations where it may not be reasonably practicable to demarcate a controlled area are when:

(a) a vehicle transporting a radioactive substance has broken down at the side of a road and there is free-flowing traffic along the road;

(b) the area is an upper room of a multi-storey building and extends outside a window, meaning the area cannot be accessed from outside – the area would only be demarcated inside the building;

(c) a person has been given a radioactive substance as part of a medical exposure and is placed in a room which has not been designated. In this case, a description of the extent of the controlled area around the patient would be sufficient;

(d) the conditions requiring a controlled area arise from the use of X-ray equipment for dental or veterinary radiography. The operator should be able to see anyone in the vicinity of the controlled area and quickly de-energise the X-ray equipment from the normal operating position.

359 The RPA can advise on whether a barrier or some other form of delineation is necessary to protect certain groups of individuals, eg building workers and window cleaners.

360 If work is of short duration and short term, such as using a mobile X-ray set in a hospital ward, it might not be reasonably practicable for the employer to demarcate the controlled area with barriers. In such cases, the person operating the equipment (or other suitable person such as the RPS) should be able to restrict access to the area by continuous supervision. The extent of the designated area should be clearly described in the local rules. The operator will need to see all the boundaries of the area and either prevent unauthorised access or terminate the radiation exposure if someone tries to enter the controlled area. The presence of a controlled area should be clearly signalled to people in the vicinity. Only continuous supervision of the boundaries of an area is likely to be effective in preventing access by people where barriers cannot be provided.

Demarcation of an area after an incident

361 If a radiation accident or other incident occurs which involves a radiation source, it might take some time to fully demarcate a controlled area around that source. If the accident was foreseeable, the employer should have identified the need to restrict access to the affected area and taken reasonable steps for this to be done, as part of the contingency plan prepared under regulation 13.

Warning signs for designated areas

362 Suitable warning signs must be placed around each designated controlled area. However, in the case of a broken-down vehicle, existing warning signs on the vehicle or packages will generally be sufficient. Signs should be located at each entrance to the area or, in the case of temporary barriers, at frequent intervals where they can be seen by people approaching them.

363 Warning signs are appropriate for some supervised areas. However, where the extent of the area is clearly set out in the local rules, and if the extent of the area is well understood by those who work there, it might not be appropriate to provide them.

364 All warning signs must comply with the minimum requirements set out in Parts I–VII of Schedule 1 of the Health and Safety (Safety Signs and Signals) Regulations 1996.[8] Employers can add any supplementary text or cautionary notice to the sign appropriate to their situation. Signs must give sufficient information to alert employees to the risks arising from the source (eg X-rays or risk of inhaling or ingesting radioactive contamination). This should enable employees who have received appropriate training or instruction to know what action to take on entering the area, for example to wear PPE.

365 If it is not reasonably practicable to physically demarcate a controlled area, it may not be practicable to provide warning signs. If so, one or more people must be present to give an appropriate verbal warning to anyone approaching the boundary of the controlled area.

(3) The employer who has designated an area as a controlled area must not permit any person to enter or remain in that area unless they –

(a) are a classified person who is not a classified outside worker;
(b) are a classified outside worker in respect of whom that employer has taken all reasonable steps to ensure that the person –
   (i)   is subject to individual dose assessment pursuant to regulation 22;
   (ii) has been provided with and has been trained to use any personal protective equipment that may be necessary pursuant to regulation 9(2)(c);
   (iii) has received any specific training required pursuant to regulation 15; and
   (iv) has been certified fit pursuant to regulation 25 for the work with ionising radiation which the person is to carry out; or
(c) not being a classified person, have entered or remain in the area in accordance with suitable written arrangements.

(4) The written arrangements referred to in paragraph (3)(c) must ensure that –

(a) an employee or a non-classified outside worker aged 18 years or over does not receive in any calendar year a cumulative dose of ionising radiation which would require that person to be designated as a classified person; or

Regulation 19(4)

(b)   any other person does not receive in any calendar year a dose of ionising radiation exceeding any relevant dose limit.

Guidance 19(3)–(4)

Responsibility for control of access

366 Employers who designate an area under regulation 17 are responsible for controlling access to that area. If the employer in control of the area hands control of that particular area to another employer, for example a contractor carrying out maintenance work in that area, the second employer will be responsible for controlling access. Different requirements apply to access by classified persons and non-classified persons. There are also additional requirements concerning access by those who are classified or non-classified outside workers. Anyone who enters a controlled area must have received appropriate information, instruction and training under regulation 15.

367 Physical barriers and warning signs are sufficient to restrict access to a controlled area when backed up by appropriate training and instruction, in accordance with regulation 15. In some cases, employers in control of the area will need to arrange for supervision of access points into the area to ensure that appropriate checks can be made.

Entry of classified persons

368 Under regulations 15, 22 and 25, employers of classified persons must make sure that any of their employees who enter controlled areas have suitable dose monitoring and adequate medical surveillance in place. If a classified person only has experience of work involving exposure to external radiation, the employer must consider what additional steps should be put in place before allowing that person access to controlled areas, designated on the basis of work with radioactive materials.

Entry of classified outside workers

369 Before a classified outside worker is allowed to enter the controlled area, the employer in control of that area must check:

(a)   their radiation passbook is up to date and contains the required information;
(b)   they have received the necessary training;
(c)   they have been passed fit by the relevant doctor to undertake the work with ionising radiation;
(d)   they are subject to routine dose assessment by an ADS (see regulations 15, 22 and 25).

370 The employer in control of the area and the employer of the classified outside worker must co-operate to make sure that any classified outside workers sent to the site have had sufficient training, to allow them to work safely in the controlled area. The classified worker must also be provided with any necessary PPE (see regulations 10 and 16). This co-operation may best be arranged as part of the contractual agreement between the two employers.

371 If no formal agreement exists, the employer in control of the area must ask outside workers what training they have received. If that training appears to be insufficient for the intended work then additional, specific training will be necessary before the employer can allow them to start work in the area.

Entry of non-classified persons or non-classified outside workers

372 Non-classified employees and non-classified outside workers should only be allowed conditional access to controlled areas. The employer in control of the area must set out these conditions in the written arrangements which should be included as part of the local rules. The conditions must set out the arrangements in place to restrict an employee's exposure to ionising radiation and should consider close supervision, the use of PPE, and restrictions on the type of work carried out, or the time spent in the area.

373 Non-classified employees and non-classified outside workers may require access to a controlled area, for example where:

(a)   a whole room has been designated as a controlled area but the work with ionising radiation takes place in only one small area, such as a bench or fume cupboard;

(b)   a person enters the area for a limited time to carry out a simple maintenance job, eg to repair a central heating radiator, to witness a test or to carry out an inspection;

(c)   work with ionising radiation occurs intermittently in the area.

Written arrangements for non-classified persons/outside workers

374 Written arrangements should restrict the exposure of non-classified persons (as required by regulation 9). It would not be reasonable for a person working under such arrangements to receive a dose which approaches the dose levels specified in regulation 19(4).

375 The requirements of regulation 9(6) regarding pregnant or breastfeeding women are also relevant. It is sufficient to require such women to inform the line manager for the area and to restrict the type of work that they can do there. The written arrangements should also include provision for estimating the dose likely to be received.

Entry by members of the public and employees who do not normally work with ionising radiation

376 If other employees who do not normally work with ionising radiation or members of the public need to enter a controlled area, eg visitors to hospitals and organised parties visiting licensed nuclear installations, arrangements must be put in place to restrict their doses. The employer could provide suitable dosemeters for these visitors, which would be one means of satisfying the requirements of regulation 19(4).

Entry of patients and research subjects

377 The requirement to restrict entry to a controlled area does not apply to any patient or research subject entering it for the purpose of receiving a medical exposure (see regulation 3).

(5)   A non-classified outside worker is not permitted to enter or remain in a controlled area pursuant to paragraph 3(c) unless they have been provided with personal protective equipment and training pursuant to paragraph 3(b)(ii) and (iii).

(6)   An employer who has designated an area as a controlled area must not permit a person to enter or remain in such area in accordance with written arrangements

Regulation	**19(6)**

under paragraph (3)(c) unless the employer can demonstrate, by personal dose monitoring or other suitable measurements, that the doses are restricted in accordance with paragraph (4).

Guidance	**19(6)**

Dose monitoring for non-classified persons

378 Where non-classified persons enter controlled areas, employers must put arrangements in place to demonstrate that any exposures received will not exceed the dose levels specified in regulation 19(4).

379 Where employers decide to provide individual dose monitoring to meet this requirement, it is desirable for it to be carried out by an appropriate ADS approved under regulation 36. Any employer who decides to make arrangements other than personal dosimetry for assessing doses to non-classified persons must be able to demonstrate that the measurements and assessments have been made to a satisfactory standard. Dosemeters or monitoring instruments (except those supplied by an ADS) used to demonstrate compliance with regulation 20(1) will need to be subject to adequate periodic tests, to ensure that any measurements made remain suitable.

380 When personal dosemeters are not used, a suitable measurement for the purposes of regulation 19(6) may be a measurement of the dose rate (or airborne contamination level), together with a record of the time spent in the controlled area. However, such measurements are only suitable if the dose rate (or contamination level) is known and is fairly constant, or is known not to exceed a particular value.

381 Employers should check the estimates of doses received by non-classified persons entering or working in the controlled area, to make sure the written arrangements are effective. It is advisable for employees who spend a significant amount of their time working in designated controlled areas to be provided with personal dose monitoring.

382 Other types of measurement may be suitable to demonstrate compliance with regulation 19(6). One way of demonstrating this is to issue a personal dosemeter (eg a direct-reading dosemeter) to one person in a group of visitors which stay together within the controlled area. The measurements made should be representative of the doses received by each member of the group. Personal dose monitoring should not be necessary for occasional external visitors who enter a controlled area, since the conditions for entry should ensure that the doses they receive will be very small.

Regulation	**19(7)**

(7)   An employer who has designated an area as a controlled area must, in relation to a classified outside worker, ensure that –

(a)   the classified outside worker is subject to arrangements for estimating the dose of ionising radiation received by that worker whilst in the controlled area;

(b)   as soon as is reasonably practicable after the services carried out by that classified outside worker in that controlled area are completed, an estimate of the dose received by that worker is entered into that worker's radiation passbook; and

(c)   when the radiation passbook of the classified outside worker is in the possession of that employer, the passbook is made available to that worker upon request.

Providing and recording dose estimates for classified outside workers

383 The estimate should be made using a suitable personal dosemeter (or personal air sampler in the case of internal radiation). The employer in control of the area should make an estimate of the total dose received once the work has been completed and enter this in the radiation passbook before the classified outside worker leaves the site (at least for external radiation). A quick dose estimate that tends to overstate the dose is preferable to a more accurate estimate that cannot be made for some days, or even weeks.

384 The accuracy of the estimate and complexity of the method used should be proportionate to the scale of the expected dose; gross overestimates may result in unnecessary restrictions being placed on workers.

385 Employers responsible for controlled areas must make suitable arrangements for recording an estimate of total doses received during the work in the classified outside worker's radiation passbook. An estimate can be entered at the controlled area itself. Alternatively, the passbook can be kept elsewhere on site and completed on production of a signed record of the cumulative dose estimate originating from the work area. The passbook should be updated on completion of the work, or when the outside worker leaves the site and carries out work for another employer in a controlled area on a different site, whichever is sooner. Estimates do not have to be entered in passbooks each time classified outside workers physically leave the site (eg at the end of a day's work).

386 Employees who record dose estimates in radiation passbooks must have had suitable training (see regulation 15).

387 The period covered by the estimate will often be different from the dose assessment period used by the approved dosimetry service for that person. The overall dose assessment will provide a more accurate picture of the dose received over a one-month or three-month period.

388 The employer responsible for the controlled area must make special arrangements for entering the dose estimate in the passbook if the work is to be done outside normal working hours. If the estimate cannot be entered while the classified outside worker is on site, the person's employer should be advised of the estimate as soon as possible. If an estimate cannot be made because of an ongoing investigation into the exposure of the worker, the employer responsible for the controlled area should note this in the classified outside worker's radiation passbook.

(8)   The employer who carries out the monitoring or measurements pursuant to paragraph (6) must keep the results of the monitoring or measurements referred to in that paragraph for a period of two years from the date they were recorded and must, at the request of the person to whom the monitoring or measurements relate and on reasonable notice being given, make the results available to that person.

(9)   In any case where there is a significant risk of the spread of radioactive contamination from a controlled area, the employer who has designated that area as a controlled area must make adequate arrangements to restrict, so far as is reasonably practicable, the spread of such contamination.

Providing information to non-classified persons

389 Where appropriate, employers must make specific arrangements for the access and exit of people and goods, and for monitoring contamination within the controlled and surrounding areas.

Guidance 19(8)–(9)

390 This provision only relates to the risk of contamination (see definition in regulation 2(1)) spreading outside controlled areas (eg from handling radioactive substances not in sealed form) leading to possible exposure to internal radiation. The risk is significant if:

(a) the employer requires employees who enter the area to wear PPE to protect them against surface or airborne contamination; or

(b) the spread of contamination would cause the employer to extend the restrictions on entry to a wider area, or to require the use of particular procedures (including the use of PPE) outside the designated area.

Regulation 19(10)

(10) Without prejudice to the generality of paragraph (9), the arrangements required by that paragraph must, where appropriate, include –

(a) the provision of suitable and sufficient washing and changing facilities for persons who enter or leave any controlled or supervised area;

(b) the proper maintenance of such washing and changing facilities;

(c) the prohibition of eating, drinking or smoking or any similar activity likely to result in the ingestion, inhalation or absorption of a radioactive substance by any employee or outside worker in a controlled area; and

(d) the means for monitoring contamination –

(i) within a controlled area and, where appropriate, in the adjacent area; and

(ii) on any person, article or goods leaving a controlled area.

Guidance 19(10)

391 This requirement is in addition to the requirements in the Workplace (Health, Safety and Welfare) Regulations 1992 for providing washing facilities, accommodation for clothing etc. The purpose of this regulation is to ensure that any radioactive contamination is not ingested or spread to areas outside the controlled area.

Washing facilities

392 The employer responsible for the controlled area must consider likely levels of contamination arising in these areas to assess the type and extent of washing and changing facilities needed. The possibility of accidents, such as spillages, must also be taken into account in deciding whether it is appropriate to provide these facilities. Employers must provide washing facilities for places where contamination is likely. What is adequate will vary from normal washing facilities for low levels of contamination, to showers where high levels of contamination can be expected. Normally, the best position for these facilities will be next to the changing facilities.

393 Wash basins should be supplied with hot and cold water via jets or sprays which can be operated without using hands (eg foot or elbow operated). Soap and drying facilities such as disposable towels must be provided by the employer; nail brushes may also be needed. Static or roller towels are not suitable in most situations.

394 The washing facilities should be accessible but situated so that they do not themselves become contaminated.

Changing facilities

395 Changing facilities are needed where protective clothing (other than disposable gloves) has to be worn in the area. Employers must provide a system to allow protective clothing or RPE that has been worn, or any other contaminated clothing, to be left initially in the controlled area. The system must prevent the spread of contamination from protective clothing to personal clothing.

396 Where protective clothing or RPE is worn in a controlled area, the following should be provided at or outside the entrance to the area:

(a) a bench or barrier to demarcate the exit. This is so protective clothing and RPE which may be contaminated can be removed and left within the area;

(b) containers on the 'active' side of the barrier for discarded contaminated clothing;

(c) lockers on the 'clean' side of the bench or barrier for uncontaminated clothing, shoes etc;

(d) a supply of clean protective clothing on the 'clean' side if not provided elsewhere.

397 In laboratories and other facilities handling small quantities of radioactive substances, arrangements which are less onerous may be sufficient, provided that cross-contamination of clean and contaminated clothes and shoes is prevented.

Maintenance of washing and changing facilities

398 Facilities are not considered to be in an efficient state unless they are maintained in a clean condition to prevent the build-up of contamination.

Eating, drinking and smoking

399 Where there is a significant risk of people ingesting radioactive materials within the controlled area because of surface contamination, employers must make arrangements to prohibit eating, chewing, drinking etc in the area. Employees can drink from a drinking fountain located in the area if it is constructed so that the water cannot be contaminated. Local rules, or signs posted in the area, can be used to reinforce any restrictions.

400 Where it is necessary to prohibit eating, drinking or smoking in the controlled area, suitable alternative facilities must be provided for these activities to be carried out in an uncontaminated area.

Monitoring facilities

401 Employers should monitor for contamination wherever areas are designated as controlled because of the presence of unsealed radioactive substances, except where the nature of the substances makes monitoring impracticable (eg radioactive gases). Where monitoring is undertaken for contamination, employers responsible for the controlled area should decide whether portable monitors are sufficient or if installed personnel contamination monitors (eg hand and foot monitors) are required. The decision will depend on the nature of the work, the type of contamination being measured, and the number of people and articles in the area.

402 Monitors must be suitable for detecting significant contamination from the radionuclides present in the area. Employers must properly maintain and test monitors to make sure they remain in working order. Advice from the RPA may be necessary (see regulation 14 and Schedule 4).

Regulation 20 Monitoring of designated areas

(1) Every employer who designates an area as a controlled or supervised area must take such steps as are necessary (otherwise than by use of assessed doses of individuals), having regard to the nature and extent of the risks resulting from exposure to ionising radiation, to ensure that levels of ionising radiation are

Regulation 20(1)–(2)

adequately monitored for each such area and that working conditions in those areas are kept under review.

(2) Adequate monitoring referred to in paragraph (1) must include –

(a) in relation to areas designated on the basis of external radiation, measurement of dose rates (averaged over a suitable period if necessary); and

(b) in relation to areas designated on the basis of internal radiation, measurements where appropriate of air activity and surface contamination taking into account the physical and chemical states of the radioactive contamination;

ACOP   20(1)–(2)

403 Monitoring should be designed to indicate breakdowns in controls or systems and detect changes in radiation or contamination levels. Monitoring is needed both inside and outside the boundaries of controlled and supervised areas, to check their continued correct designation.

Guidance   20(1)–(2)

Responsibility for monitoring designated areas

404 Employers in control of supervised and controlled areas are responsible for ensuring monitoring is carried out in those areas. They have responsibility for deciding to designate areas under regulation 17.

Purpose of monitoring

405 The main purposes of monitoring are to:

(a) check that areas have been (and remain) correctly designated;

(b) help determine radiation levels and contamination from particular operations, so that appropriate control measures for restricting exposure can be considered;

(c) detect breakdowns in controls or systems, to indicate whether conditions are suitable for work to continue in that area;

(d) provide information on which to base estimates of personal dose for non-classified persons, outside workers and classified persons when a dose assessment could not be made by the ADS.

Adequate monitoring

406 To check whether monitoring is adequate, employers must consider:

(a) the nature and quality of the radiation and the physical and chemical state of any radioactive contamination likely to be in the area;

(b) what kinds of measurements should be made (eg surface contamination, air concentrations);

(c) where the measurements should be made;

(d) how frequently, or on what occasions, these measurements should be made (including measurements forming part of contingency arrangements);

(e) what method of measurement will be used, for example direct measurement with an instrument, collection of air samples, use of wipe samples and what type of instrument to use;

(f) who should carry out the measurements and what training they need (see guidance to regulation 15);

(g) whether employees carrying out the monitoring are familiar with the proper use of the instruments and know how to interpret and record the results correctly;

(h)   how to ensure that the equipment continues to function correctly (see regulation 20(3));
(i)   what records should be kept;
(j)   how to interpret and review the results;
(k)   the selection of safe working levels and the action to be taken if they are exceeded;
(l)   when the monitoring procedures should be reviewed.

Role of the radiation protection adviser

407   Employers must consult an RPA about the implementation of requirements for controlled and supervised areas (regulation 14 and Schedule 4). These requirements include the:

(a)   type and extent of the monitoring programme;
(b)   initial calibration and regular checking of monitoring equipment to ensure it is serviceable and correctly used.

408   For this purpose, calibration means measuring the monitoring instrument's response to known radiation fields. The RPA will be able to advise about the relevant radiation fields likely to be encountered in the designated areas, which employers need to be able to monitor adequately. The RPA will also advise on the routine checks that are necessary to ensure the instrument remains operational on a daily basis. However, employers will need a qualified person to supervise or perform the examinations and tests required on instruments used to monitor those fields (see regulation 20(4)).

(3)   *The employer upon whom a duty is imposed by paragraph (1) must provide suitable and sufficient equipment for carrying out the monitoring required by that paragraph, which equipment must –*

(a)   *be properly maintained so that it remains fit for the purpose for which it was intended; and*
(b)   *be adequately tested and examined at appropriate intervals.*

Selection, maintenance and testing of equipment used for monitoring designated areas

409   Monitoring equipment should normally be tested and thoroughly examined at least once every year. A radiation risk assessment (regulation 8) will determine whether more frequent examination is necessary.

Provision of suitable monitoring equipment

410   An adequate number of suitable instruments must be available for monitoring. In deciding the number of instruments to provide, employers must take into account that sometimes they may need to be sent away for testing or repair.

411   The suitability of monitoring equipment depends on the type, nature, intensity and energy of the radiation being monitored and the equipment's conditions of use. This process of matching monitoring equipment to the working environment in which it will be used depends on the supplier's information about its performance in that environment.

412   The RPA should be consulted on the suitability of the instruments required for the type and energy range of radiations which the employer needs to monitor. Paragraphs 420–423 give advice on the separate role of the qualified person.

<table>
<tr><td>

Guidance 20(3)

</td><td>

Maintenance of instruments

413 'Maintained' is defined in regulation 2(1). The instruments will need to be checked routinely as part of their maintenance. Routine checks include battery checks, zeroing and response to make sure the equipment is still functioning satisfactorily and has suffered no obvious damage.

414 Calibration of instruments must be discussed with the RPA. All instruments should be individually calibrated before first use and as part of the annual examination and test.

Periodic examination and testing

415 The purpose of regular thorough examinations and tests is to make sure the monitoring equipment:

(a) is in good repair, is not malfunctioning and not damaged;
(b) has not lost its calibration;
(c) is suitable for the expected duration of use until it is next thoroughly examined and tested.

416 A risk assessment, as described in the Management Regulations, will help in deciding an appropriate interval between tests and examinations for equipment used regularly. Employers must rectify any significant faults that are identified; it may be necessary to retest the calibration following any repair.

417 Most examinations and tests are likely to be carried out by organisations and individuals specialising in this work. However, there is no reason that employers cannot do their own examinations and tests, provided they have a qualified person with the necessary expertise and facilities to carry out or supervise the tests (see regulation 20(4)).

418 Proper tests of monitoring instruments make use of sources or equipment that provide a known accuracy of calibration traceable to appropriate national standards. In many cases, this will involve the use of sources with accurately determined activity, for example one obtained from a laboratory accredited by the United Kingdom Accreditation Service, or the use of equipment that has been previously calibrated to known accuracy. Authoritative guidance is available from the National Physical Laboratory measurement good practice guide.[21]

</td></tr>
</table>

<table>
<tr><td>

Regulation 20(4)

</td><td>

(4) Equipment provided pursuant to paragraph (3) will not be or remain suitable unless –

(a) the performance of the equipment has been established by adequate tests before it has first been used; and
(b) the tests and examinations carried out pursuant to paragraph (3) and sub-paragraph (a) have been carried out by or under the supervision of a suitably qualified person.

</td></tr>
</table>

<table>
<tr><td>

ACOP 20(4)

</td><td>

Monitoring and test records

419 Any records of instrument tests carried out for the purposes of regulations 20(3) and (4) should be signed by a suitably qualified person. The name and contact details of that person should be stated in the record.

</td></tr>
</table>

Qualified persons for testing equipment

420 A qualified person is someone who has the combination of training, knowledge and experience and the ability to apply these to make sure that equipment is safe to use, and does not present any risk to health. They should have full knowledge and understanding of instrumentation theory and practice, currently accepted testing standards and relevant technical guidance on testing the type of monitoring equipment used (eg the National Physical Laboratory measurement good practice guide).

421 Qualified persons may be employees of the employer who designated the area or employees of an instrument manufacturer, supplier or specialist test house. Employers could give the responsibility for developing test protocols for the monitoring equipment to the qualified person, taking into account the intended use of that equipment as stated by the employer.

422 Employers in control of the area must decide what types of test are appropriate for the circumstances in which the equipment is used. They must then advise the qualified person, who is the expert on the nature and frequency of the tests required for these instruments. The qualified person is responsible for the comprehensive and competent completion of the tests, appropriate to the working environment specified by the employer. The qualified person should make sure their clients know exactly what sort of examination and calibrations they have contracted for and any relevant limitations.

423 Employers do not need to appoint the qualified person formally in writing. However, it is important that both parties clearly understand the scope of the testing work required, for example as part of a contractual arrangement.

Testing of monitoring equipment before first use

424 Testing of instruments before they are used will usually consist of individual tests of each monitoring instrument and individual calibration. Where type testing has been carried out, selected tests are usually undertaken to make sure an instrument conforms to type. If no type-test data are available, then tests are required. This is to establish the instrument's performance for its intended use and also to determine any limitations that may make it unsuitable for certain applications. Information on the limitations of any equipment and its accuracy of calibration should be available to anyone who may use it.

425 As part of the service to employers, where manufacturers, suppliers and organisations specialising in monitoring equipment carry out the tests referred to in regulation 20(3), these should still be carried out by, or under the immediate supervision of, a qualified person. They will be able to decide on the testing required depending on the type-test data, or any other reliable information they have about the design criteria and performance characteristics of that particular type of equipment, and its intended use.

(5) The employer upon whom a duty is imposed by paragraph (1) must –

(a) make suitable records of the results of the monitoring carried out in accordance with paragraph (1) and of the tests carried out in accordance with paragraphs (3) and (4);

(b) ensure that the records of the tests carried out in accordance with paragraphs (3) and (4) are authorised by a suitably qualified person; and

(c) keep the records referred to in sub-paragraph (a), or copies of those records, for at least 2 years from the respective dates on which they were made.

<table>
<tr><td>

Regulation   20(6)

</td><td>

(6)   Suitable records of the results of the monitoring referred to in paragraph 5(a) must include –

(a)   in relation to areas designated on the basis of external radiation, an indication of the nature and quality of the radiation in question;

(b)   in relation to areas designated on the basis of internal radiation, an indication, where appropriate, of the nature and physical and chemical states of the radioactive contamination.

</td></tr>
</table>

Guidance   20(5)–(6)

Monitoring and test records

426   Any records of instrument tests carried out for the purposes of regulations 20(3) and (4) must be signed by a qualified person. The name and contact details of that person must be stated in the record. These records should identify the qualified person who conducted the tests carried out on particular monitoring equipment.

427   Suitable monitoring records should include the date, time and place of monitoring and confirm that controlled and supervised areas are correctly designated and show where levels are being approached which may require investigatory or remedial action to be taken.

428   In addition to the details mentioned in paragraphs 426 and 427, the record of the monitoring results would typically show the radiation level found (ie dose rate, air concentration or surface contamination) and, where relevant, the existing conditions in relation to the radiation level measured. However, there may be no need to keep many routine monitoring results; for example in medical radiography it is not necessary to record the dose rate every time a room is entered, but periodic reviews of dose rates around the room should be recorded.

429   Employers, in consultation with the RPA, should decide in advance what needs to be recorded (see regulation 14 and Schedule 4). The employer should also include the details of the person carrying out the monitoring and the particular equipment used. These details will be important as the employer will need to demonstrate that:

(a)   the person carrying out the monitoring had received adequate training as required by regulation 15;

(b)   the equipment used for monitoring was suitable and adequately tested as required by regulation 20(3).

430   A suitable record of equipment monitoring tests should include:

(a)   identification of the equipment;

(b)   the calibration accuracy over its range of operation for the types of radiation that it is intended to monitor;

(c)   the date of the test and the name and signature of the qualified person under whose direction it was carried out.

431   The qualified person takes responsibility for the accuracy of the information in the test certificate, or other record of the test, by signing it. However, employers should make sure instruments are suitable for their working environments. The qualified person cannot be responsible for failing to test an instrument for an appropriate range of operation if the employer does not make it clear where it will be used, ie the range of energies and types of radiation encountered in the working environment.

PART 5 Classification and monitoring of persons

Regulation 21 Designation of classified persons

Regulation 21(1)–(2)

(1)   Subject to paragraph (2), the employer must designate as classified persons those of its employees who are likely to receive an effective dose greater than 6 mSv per year or an equivalent dose greater than 15 mSv per year for the lens of the eye or greater than 150 mSv per year for the skin or the extremities and must immediately inform those employees that they have been so designated.

(2)   The employer must not designate an employee as a classified person unless –

(a)   that employee is aged 18 years or over; and
(b)   a relevant doctor has certified in the health record that that employee is fit for the work with ionising radiation which that employee is to carry out.

ACOP   21(1)–(2)

Who needs to be a classified person?

432 In deciding whether a person must be classified, the employer should take account of the potential for exposure to ionising radiation (including the possibility of accidents etc which are likely to occur) as a result of the work the individual is required to undertake.

433 An employer should designate as a classified person any employee who works with any source of ionising radiation capable of giving a dose rate where it is reasonably foreseeable an employee could receive an effective dose greater than 20 mSv, or an equivalent dose in excess of a dose limit within several minutes.

Guidance   21(1)–(2)

434 To decide if an employee needs to be a classified person an employer can take account of past doses received by the employee. However, it is not sufficient to rely on the individual's avoidance of such a dose in the past to argue that classification is unnecessary. In many cases, the reason an employer designates an employee as a classified person is that the individual carries out much of their work in controlled areas. This fact alone is not sufficient to require designation, particularly where the work with ionising radiation in the area is intermittent, or carried on in one small part of the area and in either case employees are unlikely to receive significant exposure. Employers should also consider classifying employees on the basis of potential exposure from any reasonably foreseeable radiation accident.

435 Where employees are not designated as classified persons, employers must provide suitable written arrangements to allow them to enter any controlled areas.

436 Employees must be told as soon as possible if they have been designated as classified persons, so they are aware of their responsibilities to co-operate with the employer under regulation 35. Employers must arrange medical examinations for employees (see regulation 25) and specific training (see regulation 15) before assigning them to a post as classified persons.

Guidance 21(1)–(2)	

New employees who were classified persons in their previous employment

437 When a classified person changes employment, the new employer should check that the person has been certified fit for work in the last 12 months by obtaining a copy of that certificate from the previous employer. Employers must take account of any restrictions in that certificate. If the nature of the work with ionising radiation is significantly different from that covered by the certificate, the employer should arrange for the employee to receive a medical examination (see regulation 25).

438 When the requirements for medical surveillance are satisfied, then the employee can start work as soon as training/induction is complete. However, the employer must make sure that the radiation dose received is assessed (see regulation 22).

Employees on short-term contracts

439 When taking on workers on short-term contracts who need to be designated as classified persons, employers must take special care to make sure that these employees are covered by arrangements for medical surveillance, dose assessment and dose record-keeping required by regulations 22 and 25.

Female employees

440 For women who have made a declaration to their employer that they are pregnant or who are breastfeeding (regulation 9(6)), employers must be aware of the special restrictions on working conditions that will apply.

Regulation 21(3)	

(3) The employer may cease to treat an employee as a classified person only at the end of a calendar year except where –

(a) a relevant doctor so requires; or

(b) the employee is no longer employed by the same employer in a capacity which is likely to result in significant exposure to ionising radiation during the remainder of the relevant calendar year.

ACOP 21(3)	

When can an employee cease to be treated as a classified person?

441 **Exposure is significant if the employee is likely to receive an effective dose at a rate exceeding 1 mSv per year as a result of work in the new post.**

442 **Employers should note that employees, whose exposures exceed 1 mSv per year as a result of a new post, may still need to be classified due to cumulative levels of exposure.**

Guidance 21(3)	

443 Employers can only remove designation of employees as classified persons before the end of the calendar year in special circumstances, as follows:

(a) an employee ceases employment (a termination record is required – see paragraph 468);

(b) the relevant doctor certifies in the health record that an employee should not be engaged in work with ionising radiation as a classified person (see regulation 25(5));

(c) an employee is transferred by the employer to new duties which do not involve any significant exposure to ionising radiation, for example if the employee is promoted to a supervisory job which involves no further work in a designated area. If the new post did involve significant exposure, systematic monitoring of doses, as required by regulation 22, would need to continue and the person should still be treated as a classified person.

444 In most cases, employees will only stop being classified persons if they no longer work for their current employer. Where an employer wishes to stop treating an employee as a classified person on transfer to new duties, that employer would have to be satisfied that the employee will not return to their original duties before the end of the calendar year or carry out any other work involving significant exposure.

445 If it is likely that the individual will return to work involving significant exposure to ionising radiation within the year, for example because they work for a contractor undertaking work intermittently on licensed nuclear installations, they should continue to be treated as a classified person. However, the employer should advise the approved dosimetry service of periods when that employee does not receive any exposure to ionising radiation, as no dose assessments would be required for those periods.

446 Whether or not an employee remains as a classified person, the employer must make sure that doses received over any relevant dose limitation period do not exceed the limits set out in Schedule 3.

Regulation 22 Dose assessment and recording

(1) Every employer must ensure that –

(a) in respect of each of its employees who is designated as a classified person, an assessment is made of all doses of ionising radiation received by such employee which are likely to be significant; and

(b) such assessments are recorded.

(2) For the purposes of paragraph (1), the employer must make suitable arrangements with one or more approved dosimetry service for –

(a) the making of systematic assessments of such doses by the use of suitable individual measurement for appropriate periods or, where individual measurement is inappropriate, by means of other suitable measurements; and

(b) the making and maintenance of dose records relating to each classified person.

(3) For the purposes of paragraph (2)(b), the arrangements that the employer makes with the approved dosimetry service must include requirements for that service –

(a) to keep the records made and maintained pursuant to the arrangements, or a copy of those records, until the person to whom the record relates has or would have attained the age of 75 years but in any event for at least 30 years from when the record was made;

(b) to provide the employer at appropriate intervals with suitable summaries of the maintained dose records;

(c) when and as required by the employer, to provide the employer with copies of the dose record relating to any of the employer's employees;

Regulation 22(3)

(d) when required by the employer, to make a record of the information concerning the dose assessment relating to a classified person who ceases to be an employee of the employer, and to send that record to the Executive and a copy of the record to the employer as soon as possible, and such a record is referred to in this regulation as a "termination record";

(e) within 3 months, or such longer period as the Executive may agree, of the end of each calendar year to send to the Executive summaries of all current dose records relating to that year;

(f) when required by the appropriate authority, to provide it with copies of any dose records;

(g) where a dose is estimated pursuant to regulation 23, to make an entry in a dose record and retain the summary of the information used to estimate that dose;

(h) where the employer employs a classified outside worker, to provide, where appropriate, a current radiation passbook in respect of that classified outside worker; and

(i) where the employer employs a classified outside worker who works in Northern Ireland or another member State, to maintain a continuing record of the assessment of the dose received by that classified outside worker when working in such place.

Guidance 22(1)–(3)

Engaging suitable approved dosimetry services

447 Employers who designate employees as classified persons under regulation 21(1) must engage an approved dosimetry service (ADS). The ADS will carry out any necessary dose assessments, open and maintain dose records and provide relevant information from those records.

448 Dosimetry services are approved by HSE (or a body specified by HSE) under regulation 36 for one or more of the following specific purposes:

(a) the measurement and assessment of whole-body or part-body doses arising from external radiation (notably X-rays, gamma rays, beta particles or neutrons);

(b) the assessment of doses from intakes of specified classes of radionuclides;

(c) the assessment of doses following an accident or other incident (see regulation 24);

(d) the co-ordination of individual dose assessments by other approved services; making, maintaining and keeping dose records; and the provision of summary information.

449 An organisation may hold certificates of approval for more than one of these functions. However, that organisation will only carry out those approved functions covered by contractual or other formal arrangements with the employer.

450 HSE publishes criteria for approval which must be met by a dosimetry service seeking approval or wishing to remain approved. The aim of approval is to make sure, as far as possible, that doses are assessed on the basis of accepted national standards and that dose records bring together all such dose assessments, helping employers check that doses are being kept as low as reasonably practicable and dose limits are not exceeded.

451 The certificate of approval issued to each ADS shows the purpose for which it has been approved and any limitations. Employers should consider consulting the RPA about the types of dosimetry service required for the assessment and

Guidance   22(1)–(3)

recording of doses received by the classified persons. They must then engage sufficient and suitable ADSs for those purposes.

Significant doses

452 When a person is subject to the requirements for dose assessment, all doses likely to be significant must be assessed (except those received as a result of natural background radiation at normal levels and the person's own medical exposure). If personal dosemeters are used for assessing a particular component of dose, they must be worn at all times while at work when there is likely to be significant occupational exposure from that component.

453 If dose assessments are made for exposure to external gamma radiation or X-rays, employers should decide whether additional components of dose need to be assessed as well. These components might include, for example, dose from neutrons or committed dose from internal radiation. The employer's decision will depend on the expected magnitude and variations in that component. Normally, effective doses from an additional component which can reasonably be expected to exceed about 1 mSv a year would be regarded as significant in this context for that component of dose. However, it may not always be appropriate to assess a particular component of dose at this level by individual measurement.

Methods of assessing individual dose

454 There are various methods of assessing personal dose. The choice of method will depend on the circumstances of individual cases, including the nature of the work and the type of ionising radiation to be measured. Examples of different methods are the use of different types of personal dosemeters to assess:

(a) the whole-body dose, skin dose and dose to the lens of the eye;
(b) the dose to the skin of the hands; and
(c) the dose to the whole body from neutron radiation.

455 In many cases, employers will only need an assessment to be made of employees' doses received from external X-rays, gamma radiation and beta particles using an appropriate type of personal dosemeter. Assessment of committed doses arising from intakes of radionuclides into the body (by inhalation, ingestion or via wounds) is generally more difficult and may involve a combination of techniques. These include:

(a) personal air samplers;
(b) biological samples (particularly urine);
(c) monitoring of part of the body (eg the thyroid in the case of radioactive iodine) or the whole body with sensitive detectors to detect radioactive material within the body.

Appropriate dose assessment periods

456 Often the dose assessment period (eg the issue period for dosemeters) will be one month, but periods as long as three months may be appropriate where doses are very low. For external radiation, the choice of assessment period will depend on the:

(a) dose rates to which people are exposed;
(b) ability of the dosemeter to measure low doses;
(c) level of the expected dose;
(d) stability of the stored image or signal on the dosemeter over time.

Guidance 22(1)–(3)

457 Assessments must be carried out more frequently where there is a significant risk of accidental exposure, eg when working with large sealed sources outside a shielded enclosure. In this case, the use of direct reading alarming dosemeters will also be of use.

Role of the employer

458 Employers are responsible for putting the arrangements agreed with the ADS into effect. These should include:

(a) issuing dosemeters, air samplers etc;
(b) the secure storage of dosemeters or samples;
(c) the collection and despatch of dosemeters or samples to the ADS;
(d) adequate arrangements to protect dosemeters from accidental exposure when not being worn, for example by avoiding the use of X-ray security machines for dosemeters sent to the employer by the ADS.

459 The ADS must provide the employer with guidance on the general arrangements for the issuing, collecting and secure storage of dosemeters. The RPA will also be able to assist with such arrangements.

Role of employees

460 The general duties of HSWA section 7 apply to all employees, both classified and non-classified. Employees must take reasonable care of their own health and safety and co-operate with their employer to enable that employer to meet their legal obligations. In this context, this means that employees must follow their employer's instructions on the use and wearing of personal dosemeters and take reasonable care of them. Failure to do so is likely to be an offence under this section of the HSWA.

Keeping classified persons informed about the arrangements

461 Employers must make sure that classified persons have adequate training, information and instruction relating to the care and proper use of dosemeters, and the arrangements they must comply with for the issue and return of dosemeters (regulation 15).

462 Classified persons need to be made aware of their duties under regulation 35(4) to comply with reasonable requirements set out by their employer for making measurements and assessments of the doses they receive, and of their right to see copies of dose summaries etc as provided under regulation 22(6).

463 The Data Protection Act 1998[22] requires that data subjects (classified persons in this case) are made aware of the nature of the information held on them. The information provided to classified persons under IRR17 might include details of the type of data held by the ADS and HSE's Central Index of Dose Information (CIDI),[23] as provided by this regulation.

Keeping dose records

464 A formal dose record must be maintained for each classified person by a dosimetry service approved for that specific purpose (regulation 22(3)). This is to make sure that all assessed doses are properly recorded and summed for the calendar year (or other dose limitation period). The ADS responsible for maintaining the dose record will also help to obtain a dose history for the person, either from the previous ADS or from HSE's CIDI.

465 Occasionally, employers may receive reports of doses to their employees directly from the ADS which assessed those doses. This situation is only likely to arise where the ADS is unsure whether these employees are classified persons. If the employees are classified persons, the employer should pass these reports of doses to the dosimetry service approved for record-keeping; only the records kept by this service would meet the requirement under regulation 22(3).

Dose summaries

466 The main purpose of dose summaries is to help employers check that doses are being adequately restricted, eg they will show significant differences between doses received by individuals carrying out similar work. Typically, dose summaries will be provided at the same frequency as dose assessment periods. Employers should make sure that any arrangements made with the ADS for providing dose summaries suit their particular circumstances. Intervals greater than three months are unlikely to be suitable.

467 Employers will also want to know as soon as possible when the cumulative dose for a classified person is approaching, or has exceeded a relevant dose limit. The arrangements made with the ADS responsible for maintaining dose records would normally provide for this prompt information to the employer.

Termination records

468 The main purpose of termination records is to provide the new employer with relevant dose information when a classified person changes employment. It also provides a summary for the employer and the employee of the doses received during employment as a classified person with that employer.

469 In some cases, an employee might cease to be a classified person but continue to be employed by the same employer in a different post for some years. Employers must make sure that a termination record has been issued when that employee changes employer or retires. In such cases, the employer may prefer the ADS to issue a termination record immediately. A replacement termination record is needed if that employee resumes work as a classified person before ceasing employment with the employer.

Provision of radiation passbooks for classified outside workers

470 The radiation passbook, approved by HSE under these Regulations, is a document personal to the classified outside worker and cannot be transferred to other employees.

471 New passbooks can only be obtained from an ADS approved for the co-ordination of dose assessments and maintenance of dose records. The ADS can only purchase passbooks from HSE. Where appropriate, employers of classified outside workers should arrange for their ADS to allocate passbooks on their behalf. Employers may ask the ADS to issue passbooks directly to them for onward allocation to the classified outside workers.

472 A passbook approved under these Regulations can be transferred with a classified outside worker when they change employer. It can be used by the new employer until it is full, provided relevant details, such as the new employer's name and address, are entered. If the classified outside worker has left the radiation passbook with the previous employer, or the passbook is full, the new employer will need to obtain a further passbook from the ADS responsible for record-keeping.

Guidance 22(1)–(3)

Continued dose assessment for classified outside workers

473 Regulation 22(3)(i) extends an employer's duty for arranging statutory dose assessment to cover services carried out for another employer in Northern Ireland or a member state. Where the period of work outside Great Britain is less than a dose assessment period, this could be done by requiring the classified outside worker to continue wearing whatever personal dosemeter is appropriate and/or normally worn, providing the classified outside worker will not receive any significant exposure to new components of dose (eg intakes of radionuclides, neutron dose or dose to the hands, or lens of the eye) which are not accounted for/measured by that dosemeter.

474 When the period of work is longer than a dose assessment period, employers should arrange for the ADS to send replacement personal dosemeters to the classified outside worker as necessary and for these to be returned to the ADS for assessment. Alternatively, employers may choose to make arrangements (for example by contract with the other employer) for the dose assessment(s) to be made and notified by a local dosimetry service that has been recognised for such purposes by the relevant competent authority in the UK or, where appropriate, Northern Ireland or a member state. The arrangements made should ensure that any such dose assessment for an outside worker is included in that person's dose record in Great Britain.

Dose assessment and record-keeping for employees working outside the UK

475 If employees only work with ionising radiations in another member state or elsewhere outside the UK, there is no duty under these Regulations to make arrangements for dose assessment and maintenance of dose records for those individuals. Instead, employers should comply with relevant national legislation.

Regulation 22(4)

(4) The employer must provide the approved dosimetry service with such information concerning its employees as is necessary for the approved dosimetry service to comply with the arrangements made for the purposes of paragraph (2).

Guidance 22(4)

Information for the ADS

476 The effectiveness of the arrangements with the ADS will depend on information provided by the employer about each classified person. Employers must provide accurate and timely information to the ADS responsible for co-ordination and dose record-keeping about existing employees. Additionally, this information will be required for new employees whom the employer intends to designate as classified persons. This will allow the ADS to set up a new dose record and to obtain a dose history for that person. The ADS will need to know such details as:

(a) employee's full name;
(b) national insurance number;
(c) date of birth;
(d) gender;
(e) starting date of employment as a classified person with that employer;
(f) the type of work that person will be carrying out;
(g) any changes to these details.

477 Employers must provide the ADS responsible for assessing doses with certain information at the end of each dose assessment period, eg details about which classified persons wore specified dosemeters for that dose assessment period.

The ADS will also need to know as early as practicable whether any classified persons are suspected of receiving an accidental exposure or overexposure.

478 Employers should inform the ADS responsible for record-keeping about any dose records or summaries of doses they hold for employees they intend to designate as classified persons.

Individuals who have more than one employer concurrently

479 Where an employee has more than one employer at the same time (or carries out work with ionising radiation as a self-employed person), each employer will need to establish the total dose received by the worker (see also regulation 16). This is necessary to make sure that dose limits or investigation levels have not been exceeded.

480 The ADS responsible for maintaining dose records for the relevant employer will need to be informed of any information the employer has on concurrent dose records established for the individual under arrangements made by another employer. Co-operation by the ADSs on the dose information available for that employee should be facilitated by employers.

481 Each employer may choose to give a separate dosemeter to their employee. Another option would be to give the employee an approved electronic personal dosemeter to record the individual dose received while carrying out work for each employer. Employers must co-operate and share information about doses received so that they know what dose is recorded for work with each employer (see regulation 16).

Individuals on short-term contracts

482 Employers of classified workers on short-term contracts will need to make sure that the ADS responsible for opening and maintaining dose records has sufficient information about the doses received by those individuals for the calendar year to date.

483 Where the individual is already subject to a five-year effective dose limit under regulation 12(2) and Schedule 3, or needs to be made subject to this limit, information may be required for the previous four years.

484 Employers may obtain the information from a copy of the termination record provided to the individual by the former employer. If this information is not available, the employer's ADS may need to exchange information with the former ADS.

485 HSE maintains a system related to classified persons showing which ADS is currently responsible for maintaining dose records.

486 In each case, the results of dose assessments for these short-term classified persons must be entered into their dose record, even when their contract is shorter than the usual dose assessment period.

487 The employer will register any new employee with the ADS for dose record-keeping before returning any personal dosemeters worn by that person to the ADS for assessment. This applies even if the same employee is taken on repeatedly for a series of short, but separate contracts.

Regulation	**22(5)**

(5)   An employer must –

(a)   ensure that each classified outside worker employed by it is provided with a current individual radiation passbook which must not be transferable to any other worker and in which must be entered the particulars set out in Schedule 5; and

(b)   make suitable arrangements to ensure that the particulars entered in the radiation passbook are kept up-to-date during the period of employment of the classified outside worker by that employer.

ACOP	**22(5)**

Keeping classified outside worker's radiation passbooks up to date

488 Entries in passbooks should only be made by people who have been authorised by the ADS or the appropriate employer to make such entries.

Guidance	**22(5)**

489 Employers of classified outside workers must make sure radiation passbooks are valid and that suitable arrangements are in place for keeping the passbooks up to date. These arrangements should include written instructions, specifying who does what and when, unless this would be inappropriate in the circumstances.

490 Passbooks can continue to be used when a classified person changes employer. The new employer should enter the relevant details, including the employer's name and address, in the passbook. If the classified person has left the radiation passbook with the previous employer, or the passbook is full, the new employer will need to obtain a fresh passbook from the ADS responsible for record-keeping.

491 If the employer is based in an EU member state which does not provide an authorised passbook, contracts may specify that the employer should obtain one from a dosimetry service approved in Great Britain for record-keeping. The employer is responsible for making sure it contains the necessary information.

492 Radiation passbooks can be used until they are full. However, closing passbooks at the end of a calendar year before they are full can avoid the need to transfer cumulative dose information for a part-year to any replacement passbook.

493 Employers of classified outside workers should make sure that the details in the passbook, such as the date and result of the last medical review and the cumulative dose assessment for the year so far, are brought up to date as far as possible. This should be done before the classified outside worker carries out new services in the controlled area of another employer.

Regulation	**22(6)**

(6)   The employer must –

(a)   at the request of a classified person employed by the employer (or of a person formerly employed by the employer as a classified person) and on reasonable notice being given, obtain (where necessary) from the approved dosimetry service and make available to that person –

(i)   a copy of the dose summary provided for the purpose of paragraph (3)(b) relating to that person and made within a period of 2 years preceding the request; and

(ii)   a copy of the dose record of that person; and

(b)   when a classified person ceases to be employed by the employer, take all reasonable steps to provide to that person a copy of their termination record.

Guidance 22(6)

Providing dose information to classified persons

494 Classified persons can obtain personal dose monitoring information from their employer. Many employers will provide copies of dose summaries, or an extract of such information, to their employees without receiving a request, to show their commitment to keeping doses as low as reasonably practicable.

Regulation 22(7)–(8)

(7) The employer must keep a copy of the summary of the dose record received from the approved dosimetry service for at least 2 years from the end of the calendar year to which the summary relates.

(8) In this regulation, "appropriate authority" means –

(a) in connection with the application of this regulation in relation to, or in relation to any activity carried out on, any nuclear premises, the ONR;

(b) otherwise, the Executive.

Regulation 23 Estimated and notional doses and special entries

Regulation 23(1)–(2)

(1) Where a dosemeter or other device is used to make any individual measurement under regulation 22(2) and that dosemeter or device is lost, damaged or destroyed or it is not practicable to assess the dose received by a classified person over any period, the employer must –

(a) make an adequate investigation of the circumstances of the case with a view to estimating the dose received by that person during that period and either –

(i) in a case where there is adequate information to estimate the dose received by that person, send to the approved dosimetry service an adequate summary of the information used to estimate that dose and arrange for the approved dosimetry service to enter the estimated dose in the dose record of that person; or

(ii) in a case where there is inadequate information to estimate the dose received by the classified person, arrange for the approved dosimetry service to enter a notional dose in the dose record of that person which must be the proportion of the total annual dose limit for the relevant period; and

(b) in either case referred to in sub-paragraph (a), take reasonable steps to inform the classified person of the entry in their dose record and arrange for the approved dosimetry service to identify that entry as an estimated dose or a notional dose as the case may be.

(2) The employer must, at the request of the classified person (or a person formerly employed by that employer as a classified person) to whom the investigation made under paragraph (1) relates and on reasonable notice being given, make available to that person a copy of the summary sent to the approved dosimetry service under paragraph (1)(a).

Guidance 23(1)–(2)

Lost or destroyed dosemeters etc

495 Occasionally dosemeters used to assess individual doses for classified persons may be lost or destroyed. Sometimes, it may not be possible to assess the dose received by a person because the dosemeter is severely damaged or known to be accidentally or maliciously exposed to ionising radiation when not being worn (eg in X-ray security machines). Employers should provide clear

Guidance 23(1)–(2)

instructions to employees about what to do if their dosemeter is accidentally exposed, eg dropped in a radiography enclosure when exposures are taking place. In these cases, employers must tell the ADS responsible for the relevant dose assessment as soon as they are informed of the problem and carry out an investigation. Employers should consult the RPA about this investigation (see regulation 14).

Nature of the employer's investigation

496 An investigation should establish both the cause of the loss or damage to the dosemeter to prevent a recurrence, and an adequate estimate of the dose received by the individual. Employers will find it relatively easy to establish an estimate where the person was wearing an additional dosemeter or a direct reading device. In other cases, it may be possible to make an adequate estimate of the person's dose by considering information such as their pattern of work during the dose assessment period, together with any relevant monitoring data collected in accordance with regulation 20(1).

497 If the dosemeter was damaged, or exposed when not being worn, the ADS responsible for routine dose assessments may be able to assist the investigation by examining and reporting on the dosemeter separately from those returned by other employees. The employer should tell the ADS that an investigation is under way.

498 If the investigation provides adequate evidence for an estimate of the dose received by the classified person, the employer should arrange for the ADS responsible for maintaining the dose record to insert that estimate in the record. If it is not possible to make an adequate estimate of the dose received, the employer should arrange for the ADS to insert a 'notional dose' in the record, but a notional dose should only be used as a last resort. The employee concerned should also be informed.

Providing information to the approved dosimetry service

499 Where an estimate is made, employers should also make an adequate summary of the evidence used to produce that estimate (for example monitoring data for designated areas), and provide both to the ADS responsible for maintaining the dose record.

Regulation 23(3)–(4)

(3)   Subject to paragraphs (5) and (8), where an employer has reasonable cause to believe that the dose received by a classified person is much greater or much less than that shown in the relevant entry of the dose record, the employer must make an adequate investigation of the circumstances of the exposure of that person to ionising radiation and, if that investigation confirms the employer's belief, the employer must, where there is adequate information to estimate the dose received by the classified person –

(a)   send to the approved dosimetry service an adequate summary of the information used to estimate that dose;

(b)   arrange for the approved dosimetry service to enter that estimated dose in the dose record of that person and for the approved dosimetry service to identify the estimated dose in the dose record as a special entry; and

(c)   notify the classified person accordingly.

(4)   The employer must make a report of any investigation carried out under paragraph (3) and must preserve a copy of that report for a period of 2 years from the date it was made.

Regulation 23(5)–(9)

(5)   Paragraph (3) does not apply –

  (a)   in respect of a classified person subject only to an annual dose limit, more than 12 months after the original entry was made in the record; and

  (b)   in any other case, more than 5 years after the original entry was made in the record.

(6)   Where a classified person is aggrieved by a decision to replace a recorded dose by an estimated dose pursuant to paragraph (3) that person may, by an application in writing to the appropriate authority made within 3 months of the date on which that person was notified of the decision, apply for that decision to be reviewed.

(7)   Where the appropriate authority concludes (whether as a result of a review carried out pursuant to paragraph (6) or otherwise) that –

  (a)   there is reasonable cause to believe the investigation carried out pursuant to paragraph (3) was inadequate; or

  (b)   a reasonable estimated dose has not been established,

the employer must, if so directed by the appropriate authority, require the approved dosimetry service to re-instate the original entry in the dose record.

(8)   The employer must not, without the consent of the appropriate authority, require the approved dosimetry service to enter an estimated dose in the dose record in any case where –

  (a)   the cumulative recorded effective dose is 20 mSv or more in one calendar year; or

  (b)   the cumulative recorded equivalent dose for the calendar year exceeds a relevant dose limit.

(9)   In this regulation "appropriate authority" means –

  (a)   in relation to a classified person employed wholly or mainly on nuclear premises, the ONR;

  (b)   otherwise, the Executive.

ACOP   23(3)–(9)

Estimated and notional doses

500 An estimate of the dose received should be regarded as much greater than, or much less than, the original entry in the dose record for a particular period where:

  (a)   for recorded doses of 1 mSv or less, the dose received differs from the original entry in the dose record by at least 1 mSv; or

  (b)   for recorded doses in excess of 1 mSv, but less than the relevant dose limit, the dose received differs from the original entry in the dose record by a factor of 2 or more; or

  (c)   for recorded doses above the relevant dose limit, the dose received differs from the original entry in the dose record by a factor of 1.5 or more.

Guidance   23(3)–(9)

501 Employers must carry out an investigation if there is a reason to believe that the dose recorded in a classified person's dose record is substantially incorrect. For example, the employer may discover that a classified person has not been wearing a dosemeter at certain times when working with ionising radiation. Conversely, the employer may find that a batch of dosemeters was inadvertently exposed to ionising radiation during transit or storage.

502 The dose record entry might relate to an assessment by the ADS of whole-body or part-body dose, the neutron component of dose, or committed dose from intakes of radionuclides. Any change to a single entry will affect cumulative totals in recorded doses for the calendar year so far. The appropriate authority's consent is required in certain cases where the cumulative total exceeds a certain level (regulation 23(8)).

503 Further advice is given in an HSE information document on special entries[24] which has been made available to approved dosimetry services and can be downloaded from HSE's website.

Deciding whether doses are much greater than or much less than the recorded dose

504 The ACOP guidance in paragraph 500 explains the circumstances where an investigation would be needed. There may be other circumstances where the employer considers that the dose received was much greater or much less than that recorded in the dose record, for example where a series of dose entries for consecutive dose assessment periods is affected. The employer will have to follow the procedure in regulation 23(3) in these cases, subject to the restrictions in regulations 23(5) and (8).

Unassessed components of dose

505 The provision for making alterations to recorded doses (special entries) only applies where dose assessments have already been recorded in the dose record. If the employer finds that a component of dose (eg one to the hand) not previously assessed by an ADS is, in fact, significant, estimates of this new (and unrecorded) component may be added to the dose record by the ADS responsible for maintaining that record. The requirements of regulation 23(3) will not apply as none of the existing entries in the dose record relating to assessed doses needs to be changed.

Nature of the investigation

506 The employer's investigation into the circumstances of the exposure should take account of:

(a) relevant information provided by the ADS including the possibility of defects in or contamination of dosemeters;

(b) details of the individual's pattern of work, such as the time spent in particular controlled and supervised areas;

(c) measurements from any additional dosemeter or direct reading device worn by the person concerned;

(d) individual measurements made on other employees carrying out the same work with ionising radiation;

(e) the results of monitoring for controlled and supervised areas carried out in accordance with regulation 20;

(f) details of the work routine of immediate work colleagues, as well as the individual, during the period concerned;

(g) details of any radiation monitors/personal alarms in use in the areas concerned during the period under investigation (including type, periods of use and calibration arrangements);

(h) results of any special radiation survey in the areas concerned (eg as part of a reconstruction advised by the RPA);

(i) arrangements for security and storage of the source(s) of ionising radiation;

(j) arrangements for storage/security of dosemeters or biological samples.

507 The information used to estimate the dose received will be adequate if it:

(a) shows that there is reasonable cause to believe that the dose received by the classified person was much greater than or much less than the dose recorded in the dose record;

(b) includes sufficient information to permit a reliable reconstruction of the exposure conditions for the person during the relevant dose assessment period.

508 The investigation report should at least include the information in paragraph 507 (a) and (b).

509 Employers should normally consult the RPA about this investigation. It may be appropriate to consult the appointed safety representative(s) (or established safety committee) and also the relevant RPS.

510 Where an estimate is made, the employer will also need to provide an adequate summary of the evidence used to produce that estimate.

Report of investigation

511 Employers should ensure that the report of the investigation at least includes the reasons for confirming the belief that the recorded dose is incorrect, and the basis for any estimate of dose to replace the entry in the dose record. The employer would normally make the report (or the summary provided to the ADS) available to the established safety committee or to the appointed safety representative(s), as well as to the person affected.

Restrictions on arranging special entries

512 Employers may ask the ADS to make a special entry replacing the original entry in the dose record for a classified person, only if:

(a) an adequate investigation has been conducted which allows the employer to make a sufficient estimate of the dose received during the relevant assessment period;

(b) the original entry was made within the period specified in regulation 23(5).

513 If the cumulative recorded dose for that employee in the calendar year exceeds the levels set out in regulation 23(8), the employer will need the appropriate authority's consent before making arrangements with the ADS for a special entry in the dose record.

514 The ADS responsible for maintaining the dose record will identify the replacement dose as a special entry in the record. This will flag such special entries in the annual dose summaries sent to HSE's CIDI.[23] The appropriate authority may decide to review any such entries and, where appropriate, take action under regulation 23(7) to direct the employer to arrange for the original dose entry to be reinstated.

Informing the classified person of the outcome of the investigation

515 Normally, an employee would be involved in some way in the investigation and would be informed if the employer decided to replace a recorded dose entry with an estimated dose. Employees should have the opportunity to tell their employer anything relevant to the investigation and, subsequently, to consider

Guidance   23(3)–(9)

applying to the appropriate authority for a review of the employer's decision if aggrieved by that decision.

Appropriate authority review of proposed special entries

516   The appropriate authorities may review the evidence for a special entry where:

(a)   its consent is required under regulation 23(8);

(b)   an employee applies for a review of the employer's decision to arrange for a special entry to be entered in the dose record;

(c)   it suspects that an adequate investigation was not carried out by the employer.

517   Where consent is required under regulation 23(8), employers must send the appropriate authority sufficient evidence to support the arguments for a special entry in the dose record.

518   If the appropriate authority is not satisfied with the employer's investigation or the basis for the estimated dose, it may withhold consent to a special entry or direct the employer to arrange for the ADS to reinstate the original entry in the dose record.

Regulation 24 Dosimetry for accidents etc

Regulation 24(1)–(3)

(1)   Where any accident or other occurrence takes place which is likely to result in a person receiving an effective dose of ionising radiation greater than 6 mSv or an equivalent dose greater than 15 mSv for the lens of an eye or greater than 150 mSv for the skin or the extremities, the employer must –

(a)   in the case of a classified person, arrange for a dose assessment to be made by the approved dosimetry service as soon as possible;

(b)   in the case of an employee to whom a dosemeter or other device has been issued in accordance with regulation 13(2), arrange for that dosemeter or device to be examined and for the dose received to be assessed by the approved dosimetry service as soon as possible;

(c)   in any other case, arrange for the dose to be assessed by an appropriate means as soon as possible, having regard to the advice of the radiation protection adviser.

(2)   In each such case, the employer must –

(a)   take all reasonably practicable steps to inform each person for whom a dose assessment has been made of the result of that assessment;

(b)   notify the appropriate authority of the result of the dose assessment as soon as possible; and

(c)   keep a record of the assessment until the person to whom the record relates has or would have attained the age of 75 years but in any event for at least 30 years from the date of the relevant accident.

(3)   In this regulation "appropriate authority" means –

(a)   in relation to an accident or other occurrence as a result of work carried out on nuclear premises, the ONR;

(b)   otherwise, the Executive.

General

519 In most cases, people likely to be involved in incidents of the type covered by the regulation will either be:

(a) classified persons who are subject to systematic dose assessments under regulation 22(2);

(b) employees who have been issued with a dosemeter or other device (eg criticality dosemeter) provided by an ADS under regulation 13(2).

520 In general, dosimetry services approved for the assessment of doses from external radiation are capable of providing a prompt assessment of dose, on request, in the event of exposure to ionising radiation arising from accidents or other occurrences.

521 In special cases where life-threatening doses might be received, a few dosimetry services have been specially approved for assessing doses from external radiation which may exceed 0.5 gray. In these situations it is vital to provide early dose information to medical officers about those people who may have received an exposure which could be life-threatening.

522 When an employer becomes aware that an exposure may have occurred as a result of an accident or other incident, it is important that they advise the ADS responsible for assessing doses as early as practicable. The employer will then need to arrange for dosemeters, other devices or bio-assay samples to be dispatched, without undue delay, to the ADS for analysis. If doses around 0.5 gray or above are suspected, time will normally be critical. Where appropriate, the employer should arrange for the ADS to identify the dose as an accidental dose in the dose record (see regulation 26).

523 In cases where regulation 24(1)(c) applies, appropriate means for the assessment of dose may include:

(a) examination of biological specimens, eg hair, nail clippings, blood etc;

(b) computation of dose from measured dose rates or contamination levels, together with a knowledge of exposure times in the area and distance from the place of measurement, depending on the advice of the RPA.

524 The employer should seek advice from the RPA about the investigation and, where appropriate, the relevant doctor who undertakes medical surveillance in accordance with regulation 25(2).

Application to medical exposures

525 This regulation does not apply to the protection of those undergoing a medical examination or treatment.

Regulation 25 Medical surveillance

(1) This regulation applies in relation to –

(a) classified persons and persons whom an employer intends to designate as classified persons;

(b) employees who have received an overexposure and are not classified persons;

(c) employees in respect of whom a relevant doctor has made a certification under paragraph (5).

<table>
<tr><td>

Guidance   25(1)

</td><td>

When is medical surveillance required?

526 Employers of anyone exposed to ionising radiations as a result of work activities must decide when such an employee needs to be designated as a classified person (see regulation 21). Before that person is classified, employers must make sure that the employee has been certified as fit for the intended type of work within the previous 12 months. This may require a medical examination. Employers will then need to make arrangements with the relevant doctor for continuing medical surveillance (see regulation 25(3)). Also, employers must arrange for adequate medical surveillance for any employee who has received an overexposure, whether or not that employee has been designated as a classified person (see also guidance to regulation 26).

Relevant doctor

527 The term relevant doctor is defined in regulation 2(1) and means an appointed doctor or an employment medical adviser.

</td></tr>
<tr><td>

Regulation   25(2)

</td><td>

(2)   An employer must ensure that –

(a)   each of its employees to whom this regulation relates is under adequate medical surveillance by a relevant doctor for the purpose of determining the fitness of each employee for the work with ionising radiation which that employee is to carry out;

(b)   a health record containing the particulars referred to in Schedule 6 is made and maintained in respect of such employees; and

(c)   the record or a copy of the record is kept until the person to whom the record relates has or would have attained the age of 75 years but in any event for at least 30 years from the date of the last entry made in it.

</td></tr>
<tr><td>

Guidance   25(2)

</td><td>

Purpose of medical surveillance

528 The purpose of medical surveillance is to confirm that an individual is fit or continues to be fit for the intended work with ionising radiation. Fitness of the person is not restricted to possible health effects from exposure to ionising radiation. The relevant doctor must consider specific features of the work with ionising radiation and the fitness of the individual where appropriate:

(a)   to wear any PPE (including RPE) required to restrict exposure;

(b)   with a skin disease, to undertake work involving unsealed radioactive materials;

(c)   with psychiatric illness or personality disorder, to undertake work with radiation sources that involve a special level of responsibility for safety;

(d)   with a history of chronic pulmonary disease, blood disorder, treatment with cytotoxic drugs, inherited predisposition to cancers, or previous significant medical exposure to ionising radiation.

529 The relevant doctor's decision on fitness will be made on an individual basis and the above considerations will not automatically exclude an individual from classified work.

530 In some posts, the nature of the work may be such that an employee is at risk from acute exposure to high levels of external radiation (eg site radiography) as a result of an accident. In such cases, the relevant doctor may take into account this potential for acute exposures to ionising radiation in deciding what level of medical surveillance is appropriate, even where recorded doses are generally low.

</td></tr>
</table>

Guidance   25(2)

531 For workers who have received an overexposure, medical surveillance (including monitoring for possible biological effects) is mainly intended to assess fitness to continue the work with ionising radiation. It also gives the worker the opportunity to discuss any concerns that they may have in relation to the risks resulting from the overexposure.

Medical examination before designation as a classified person

532 In general, a medical examination will not usually be needed when a classified person changes employment if the person has been certified fit for that type of work with ionising radiation within the preceding 12 months. A copy of that certification should be obtained from the previous employer and kept in the health record. The relevant doctor may also obtain previous clinical information, with the co-operation of the doctor who undertook the most recent medical surveillance for the individual. Any conditions on the individual's work already imposed by the previous doctor would continue at least until the next periodic review.

533 Where there is a change of work which will involve exposure to a different risk from ionising radiation, the relevant doctor may decide that a medical examination is necessary to determine whether the person is fit for work in the new duties. This may be considered, for example if the employee changes duties involving work with sealed sources to work with unsealed radionuclides, or has to use PPE for the first time.

Periodic reviews of health

534 After the initial medical examination conducted before designation as a classified person (regulation 25(2) and regulation 25(3)), periodic reviews of health should take place at least once every year. The relevant doctor may specify a shorter period between reviews.

535 The format of the review is decided by the relevant doctor who will take into account any guidance issued to relevant doctors by HSE. The review will involve at least an assessment of the dose profile for the individual and sickness absence records. The relevant doctor may also need access to other records concerning the working conditions of the classified person, for example records of monitoring kept in accordance with regulation 20.

536 Periodic reviews may also involve an interview with the individual and occasionally a medical examination and medical tests, depending on the nature of the work and the individual's state of health. Employers are responsible for making any necessary arrangements for medical surveillance. If the relevant doctor wishes to see the individual as part of the periodic review, the employer must arrange for that person to see the doctor at an appropriate time during working hours (see regulation 35(5)).

Special medical surveillance

537 Special medical surveillance for any employee who had received an overexposure (and is subject to an investigation under regulation 26) may be necessary. Special medical surveillance may include a medical examination if the relevant doctor considers this to be necessary in the circumstances. The medical adviser will work in consultation with the relevant doctor and others as appropriate, to determine the content of special medical surveillance. It should include a medical assessment, counselling and detailing of possible restrictions on further exposure. Specific tests, such as chromosome aberration analysis, may be warranted to help establish the degree of any overexposure. This will depend on

the size and distribution of the dose received and may not be appropriate unless a whole body exposure in excess of 100 mSv has been received.

Provision of facilities

538 Employers should either provide suitable facilities for the relevant doctor (or employment medical adviser) or allow employees to visit the doctor's surgery or examination room. Employers are responsible for the full cost, including time off for attendance.

Health record

539 Employers may use any format for the health record provided it contains at least the particulars in Schedule 6. Confidential clinical information should not be recorded in the health record but kept in a suitable medical record by the relevant doctor. The Data Protection Act 1998[22] contains data protection requirements relevant to medical surveillance records. These requirements include the right of data subjects to see their health records.

Regulation 25(3)–(4)

(3) Subject to paragraph (4), an employer must ensure that there is a valid entry made by a relevant doctor in the health record of each of its employees to whom this regulation relates (other than employees who have received an overexposure and who are not classified persons) and an entry in the health record is valid –

(a) for 12 months from the date it was made or treated as made by virtue of paragraph (4);
(b) for such shorter period as is specified in the entry by the relevant doctor; or
(c) until cancelled by a relevant doctor by a further entry in the record.

(4) For the purposes of paragraph (3)(a), a further entry in the health record of the same employee, where made not less than 11 months nor more than 13 months after the start of the current period of validity, is to be treated as if made at the end of that period.

Guidance 25(3)–(4)

Valid entries in health record for periodic reviews of health

540 Unless the relevant doctor judges they are needed more frequently in particular cases, reviews of health must take place annually. However, regulation 25(3) allows some flexibility in timing of reviews to avoid the need for entries to be made on the exact anniversary of previous ones. The review may be carried out up to one month before or one month after the due date but treated as if it had been carried out 12 months since the last review. For example, if a review which is due on 10 July is carried out between 10 June and 10 August, the next review will be due on 10 July the following year.

541 The relevant doctor should make the signed entry in the record, not the employer (see Schedule 6). The contact details of the doctor who made the entry would normally be given in the health record.

Regulation 25(5)

(5) Where a relevant doctor has certified in the health record of an employee that in their professional opinion that employee should not be engaged in work with ionising radiation or that the employee should only be so engaged under conditions specified by the relevant doctor in the health record, the employer must not permit that employee to be engaged in the work with ionising radiation, or only permit the employee to be engaged in the work in accordance with the conditions so specified, as the case may be.

Guidance 25(5)

Employees certified unfit/fit subject to conditions

542 If appropriate, the relevant doctor will list any conditions in the health record which should be followed in order to ensure the employee remains fit to continue the work with ionising radiation, including the type of work to which the conditions relate.

543 Conditions specified in the health record must be followed until such time as they are rescinded by a relevant doctor. If the relevant doctor declares that a person is unfit for certain types of work with ionising radiation, employers must not allow that person to carry out that work with ionising radiation.

544 The conditions specified in the health record may, for example, place restrictions on the type of work undertaken or the maximum dose of radiation a person should receive. The restrictions might be applied to the use of RPE or work with unsealed radioactive sources.

545 Employees can ask for a review of any decision by the doctor if they do not agree with any condition recorded in the health record or a finding that they are not fit for the work with ionising radiation.

Regulation 25(6)–(7)

(6) Where a relevant doctor requires to inspect any workplace for the purposes of carrying out their functions under these Regulations, the employer must permit them to do so.

(7) An employer must make available to the relevant doctor the summary of the dose record kept by the employer pursuant to regulation 22(7) and such other records kept for the purposes of these Regulations as the relevant doctor may reasonably require.

Guidance 25(6)–(7)

Information made available to the relevant doctor

546 The records made available to the doctor before the periodic review of health is carried out should always include any relevant records of sickness absence for the person as well as the health record and copies of the summaries of the dose record provided by the approved dosimetry service and retained in accordance with regulation 22(7).

547 The relevant doctor must have direct access to dose summary information provided to the employer by their ADS. Employers can make any agreed arrangement for the relevant doctor to see this summary. It does not need to be available as a paper record. The employer should make sure that the relevant doctor also has a copy of any explanation provided by the ADS of codes or abbreviations used in the dose summary. The health record must include the contact details of the ADS so the relevant doctor can resolve any difficulties with interpretation of the dose summary.

548 Normally, the relevant doctor would automatically see copies of sickness absence records. They may also ask to see copies of other records kept for the purposes of these Regulations. The employer should make these available to the relevant doctor if sufficient notice is given. In addition, the relevant doctor may wish to review other information concerning the specific work with ionising radiation intended to be carried out.

Regulation 25(8)

(8) Where an employee is aggrieved by a decision recorded in the health record by a relevant doctor the employee may, by an application in writing to the Executive made within 28 days of the date on which the employee was notified of the decision, apply for that decision to be reviewed in accordance with a procedure

Regulation 25(8)

approved for the purposes of this paragraph by the Executive, and the result of that review must be notified to the employee and entered in the employee's health record in accordance with the approved procedure.

Regulation 26 Investigation and notification of overexposure

Regulation 26(1)–(4)

(1)   Where an employer suspects or has been informed that any person is likely to have received an overexposure as a result of work with ionising radiation carried out by that employer, that employer must make an immediate investigation to determine whether there are circumstances which show beyond reasonable doubt that no overexposure could have occurred and, unless this is shown, the employer must –

(a)   as soon as practicable notify the suspected overexposure to –
(i)   the appropriate authority;
(ii)   in the case of an employee of some other employer, that other employer; and
(iii)   in the case of the employer's own employee, the relevant doctor;
(b)   as soon as practicable take reasonable steps to notify the suspected overexposure to the person affected;
(c)   make or arrange for such investigation of the circumstances of the exposure and an assessment of any relevant dose received as is necessary to determine, so far as is reasonably practicable, the measures, if any, required to be taken to prevent a recurrence of such overexposure; and
(d)   immediately notify the results of the investigation and assessment referred to in sub-paragraph (c) to the persons and authorities mentioned in sub-paragraph (a) and must –
(i)   in the case of the employer's employee, immediately notify that employee of the results of the investigation and assessment; or
(ii)   in the case of a person who is not the employer's employee, where the investigation has shown that that person has received an overexposure, take all reasonable steps to notify that person of their overexposure.

(2)   An employer who makes any investigation pursuant to paragraph (1) must make a report of that investigation and must –

(a)   in respect of an immediate investigation, keep that report or a copy of the report for at least 2 years from the date on which it was made; and
(b)   in respect of an investigation made pursuant to paragraph (1)(c), keep that report or a copy of the report until the person to whom the record relates has or would have attained the age of 75 years but in any event for at least 30 years from the date on which it was made.

(3)   Where the person who received the overexposure is an employee who has a dose record, the employee's employer must arrange for the assessment of the dose received to be entered into that dose record.

(4)   In this regulation 'appropriate authority' means –

(a)   in relation to overexposure as a result of work carried out on nuclear premises, the ONR;
(b)   otherwise the Executive.

Application to medical exposures

549 As they are not subject to dose limits, this regulation does not apply to the protection of those undergoing a medical exposure. However, it does apply to staff that carry out those exposures and members of the public.

Need for an investigation

550 The regulation applies to any overexposure, or suspected overexposure. This could arise from a single incident or because the total of doses received by the person during the period in question exceeds any dose limit for that period. It also includes situations where an individual is subject to a further restriction for the remainder of the dose limitation period (regulation 27) and is suspected of receiving a dose in excess of that additional restriction.

551 Employers must investigate an overexposure to anyone who works with ionising radiation. The requirements apply equally to both classified and non-classified workers, for example, the relevant doctor must be consulted if a non-classified worker is involved in an overexposure.

Nature of immediate investigation

552 The main purpose of the immediate investigation is to rule out suspected incidents which it can readily be shown did not take place. For example, it might be suspected that an individual had been inadvertently exposed to the beam from an X-ray set, but the immediate investigation showed that the set was not in fact energised at the time. In cases where a significant exposure cannot be excluded, reviews carried out to make a more accurate estimate or assessment of the dose received are not part of the immediate investigation. Such reviews can be carried out, if appropriate, during the detailed investigation that follows.

Nature of the detailed investigation

553 A detailed investigation under regulation 26(1)(c) is carried out to establish:

(a)   why the overexposure occurred;
(b)   what dose was received by the individual;
(c)   what steps are necessary to prevent a recurrence of that overexposure.

554 In addition to the person/people affected, the employer should consult the RPA about these investigations. Employers should also seek the views of relevant safety representative(s) and any established safety committee.

555 The extent of the investigation will depend on the difficulty in establishing the cause of the overexposure and the magnitude of the dose(s) received. The ADS will need to record the assessed doses separately in the person's dose record as accidental doses.

556 To make sure that the requirements of regulation 26(1) and the considerations required by regulation 26(1)(c) are satisfied, the detailed investigation should be completed within a period of no longer than three months following the date that the overexposure is discovered.

557 If the person who received the overexposure is an employee, the relevant doctor carrying out medical surveillance for that person should be involved in the investigation as soon as is practicable. The doctor will be able to discuss any concerns resulting from the overexposure and may specify conditions for future work with ionising radiation. If necessary, the employee should be made a

Guidance 26(1)–(4)

classified person and a health record should be opened for that employee. The relevant doctor may wish to seek the views of the employee concerned and take account of any likely future exposure of that person, as advised by the employer.

558 A detailed investigation should include consideration of:

(a) the work routine of the individual, and immediate work colleagues, during the period concerned;

(b) any radiation monitors/personal alarms in use in the relevant areas during the period under investigation;

(c) any known incidents where the individual may have received an unusual exposure;

(d) assessed or estimated doses over the last few years, compared with those of work colleagues carrying out similar work;

(e) results of any special radiation survey in the relevant areas (eg as part of a reconstruction advised by the RPA) to identify any deterioration in physical control measures;

(f) adherence to local rules or deficiencies in local rules;

(g) training, instruction or information received and general competence for the work undertaken;

(h) other possible explanations for a suspected overexposure (eg there is evidence that the employee has continued to wear a dosemeter while receiving a medical exposure);

(i) the need for a reconstruction of the incident to measure the exposures that could have been received.

Report of investigation

559 Employers must make the report or summary available to the established safety committee or to appointed safety representative(s) and at least a summary to the individual employees concerned.

Regulation 27 Dose limitation for overexposed employees

Regulation 27(1)–(4)

(1) Without prejudice to other requirements of these Regulations and in particular regulation 25(5), where an employee has been subjected to an overexposure paragraph (2) applies in relation to the employment of that employee on work with ionising radiation during the remainder of the dose limitation period, where that remaining period commences at the end of the personal dose assessment period in which that employee was subjected to the overexposure.

(2) The employer must ensure that an employee to whom this regulation relates does not, during the remainder of the dose limitation period, receive a dose of ionising radiation greater than that proportion of any dose limit which is equal to the proportion that the remaining part of the dose limitation period bears to the whole of that period.

(3) The employer must inform an employee who has been subjected to an overexposure of the dose limit which is applicable to that employee for the remainder of the relevant dose limitation period.

(4) In this regulation, "dose limitation period" means, as appropriate, a calendar year or the period of five consecutive calendar years.

Guidance 27(1)–(4)

560 Employees who have received an overexposure may be allowed to continue to work with ionising radiation, provided that:

(a) the provisions of regulation 26(1) have been fully complied with, by means of a detailed investigation etc;

(b) the employee has been subject to medical surveillance for overexposed workers;

(c) the relevant doctor has been involved in the decision to assign a revised dose limit to the employee;

(d) the work is performed in accordance with any conditions imposed by the relevant doctor by an entry in the health record to ensure the person remains fit for work with ionising radiation.

561 Where an employee is subject to the five-year limit on effective dose, the employer must make sure that they do not receive more than the stated proportion of the five-year dose limit for the five-calendar-year period. This includes the calendar year in which the overexposure was received.

PART 6 Arrangements for the control of radioactive substances, articles and equipment

Regulation 28 Sealed sources and articles containing or embodying radioactive substances

Regulation	**28(1)**

(1)   Where a radioactive substance is used as a source of ionising radiation in work with ionising radiation, the employer must ensure that, whenever reasonably practicable, the substance is in the form of a sealed source.

Guidance	**28(1)**

562 The definition of a sealed source is given in regulation 2(1). Using sealed sources whenever reasonably practicable will minimise the risk of dispersal of radioactive material.

Regulation	**28(2)**

(2)   The employer must ensure that the design, construction and maintenance of any article containing or embodying a radioactive substance, including its bonding, immediate container or other mechanical protection, is such as to prevent the leakage of any radioactive substance –

(a)   in the case of a sealed source, so far as is practicable; or

(b)   in the case of any other article, so far as is reasonably practicable.

Guidance	**28(2)**

563 This regulation applies to sealed sources, high-activity sealed sources and articles containing unsealed radioactive material. Employers must make sure the design and construction of the source or article is suitable for its intended use, taking account of the actual work to be done. The radiation risk assessment required by regulation 8 must consider this.

564 Information from the manufacturers or suppliers of sources and articles about the mechanical protection used, including bonding materials, can be used by employers to assess whether the manufacturing specification suits the intended use. The priority must always be to prevent loss of containment, eg if the source is to be used in wet or aggressive conditions, the possibility of corrosion should be taken into account.

565 Where a sealed source reaches the end of the working life for the source capsule recommended by the supplier or manufacturer, its condition should be reviewed, with a view to it being replaced or examined by the supplier or manufacturer. The purpose of the examination is to determine whether it is safe to extend the working life. If the source is not to be replaced, and the examination confirms that it is still safe to use, a time limit on its continued use should be set. After this time, a further review should be carried out.

566 Where the supplier or manufacturer does not specify a recommended working life for the sealed source, a review should be carried out within five years of manufacture, or advice should be sought from the RPA about a more appropriate review period, taking account of the circumstances.

Regulation 28(3)	

(3)   The employer must –

(a)   ensure that, where appropriate, suitable tests are carried out at suitable intervals to detect leakage of radioactive substances from any article to which paragraph (2) applies; and

(b)   make a suitable record of each such test and retain that record for at least 2 years after the article is disposed of or until a further record is made following a subsequent test to that article.

ACOP   28(3)

Suitable leak tests

567 Where testing is appropriate under normal operating conditions, the interval between tests should not exceed two years.

Guidance   28(3)

568 The purpose of a leak test is to show that the mechanisms for preventing dispersal of radioactive substances are effective. The assessment required by regulation 8 must identify potential ways in which containment could be lost and the likelihood of occurrence. A test method and a frequency of testing must then be chosen that can detect leakage of radioactivity from the source or article before a radiation risk arises.

569 Leak tests should have clear pass/fail criteria and should be carried out directly on the sealed source capsule. This provides the best possible check for loss of containment. However, where this is not reasonably practicable, for instance when the source is inaccessible or when significant exposures are likely to occur from performing the test in this way, an indirect test should be carried out. This indirect test should be conducted on parts of the source containment or apparatus that can reasonably be expected to have become contaminated by a leak.

570 The decision to carry out indirect testing needs to be balanced against the radiological implications of failing to detect a loss of containment, which could itself result in significant exposures. Also, particular care is needed in dismantling, or gaining closer than normal access to a source when conducting indirect testing, due to the greater risk of contamination.

571 The manufacturer or supplier will advise about periodic leak testing and the methods to adopt to give the required assurance that radioactive material will not disperse. In the absence of such advice, test methods set out in ISO 9978:1996[25] may be appropriate.

572 Tests are not usually considered appropriate in the following circumstances:

(a)   where the sealed source contains solely gaseous radioactive substances;
(b)   on an article containing a radioactive substance which is solely designed and used for the purpose of detecting smoke or fire, and is installed in a building for that purpose;
(c)   where an article containing a radioactive substance, not being a sealed source, is by design open, for example a syringe, bottle or similar equipment;
(d)   on any sealed sources during irradiation in a nuclear reactor.

573 Leak tests may not be appropriate in other situations, eg on sources or articles which do not have dimensions greater than 5 mm, such as gold grains and microspheres. These may be treated as dispersible radioactive substances. However, this advice is not intended to stop leak tests being carried out on smaller sources where it is appropriate. Even where leak tests are not carried out, care is always needed to prevent contamination.

Guidance   28(3)

Suitable interval for leak tests

574 The interval between tests under normal operating conditions should not exceed two years. There may, however, be practical difficulties in carrying out these tests every two years, although this should be the aim. More frequent testing is required for situations where the radiological implications of a loss of containment could be severe, or the physical or chemical conditions are such that deterioration of the source or its containment might occur, eg in a hot and humid environment. Additional tests for leakage are also required if any damage is suspected, or where any work has been carried out which could have affected the structure of the capsule or article.

575 Where a source or article is inaccessible in normal operation, leak testing can sometimes wait until access is possible, eg during scheduled maintenance, rather than proceeding with the routine test; for example, when gauges are installed inside process vessels or in bunkers, where gaining access during normal operation can pose a significant physical risk. The period between leak tests on a particular source does not always need to be the same and can be varied to allow for an improved test.

576 More frequent leak testing is required when a sealed source is going to be retained in use beyond the recommended working life given to the source capsule by the supplier or manufacturer. Where there is no recommended working life, then the frequency of leak testing needs to be considered as part of the periodic reviews of its condition.

Suitable record of leak tests

577 A suitable leak test record includes the:

  (a) identification of the source or article which is the subject of the test;
  (b) date of test;
  (c) reason for test (eg pre-use, manufacturer's test, nominal routine, after incident);
  (d) methods of test, including:
    (i)   when the source or article has not been tested directly;
    (ii)   a statement of what part of the device was tested;
    (iii) a statement about whether this is likely to detect any leaking material.
  (e) a statement of the pass/fail criteria;
  (f) numerical results of the test;
  (g) result of the test (pass or fail);
  (h) any action taken if the source failed the test;
  (i) name and signature of the person carrying out the test.

Regulation 29 Accounting for radioactive substances

Regulation   29(1)

(1) Every employer, for the purpose of controlling radioactive substances which are involved in work with ionising radiation undertaken by that employer, must –

  (a) take such steps as are appropriate to account for and keep records of the quantity and location of those substances; and
  (b) keep those records or a copy of the records for at least 2 years from the date on which they were made and, in addition, for at least 2 years from the date of disposal of that radioactive substance.

ACOP 29(1)

578 The procedures for accounting should ensure that the location of radioactive substances is known so that losses or theft of significant quantities can quickly be identified. A frequency for checking the location of the source should be determined, taking account of the likely movement of the source, its potential for being displaced and its susceptibility to damage. For portable sources, such as radiography sources and portable gauges, the check should be carried out at least on each working day.

Guidance 29(1)

579 Employers registered under the Environmental Permitting (England and Wales) Regulations 2016 (EPR)[19] and for Scotland the Radioactive Substances Act 1993 (RSA)[20] may also be legally required to keep records. In these circumstances, a single system of accounting can be used which satisfies both sets of legal requirements.

Accounting – general advice

580 Source accountancy records should be sufficient to quickly identify any reasonably foreseeable loss, theft, release or spillage.

581 The records for accounting for any particular radioactive substance should include:

(a) a means of identification, which for sealed sources should usually be unique. A picture of the source or article may assist with easy identification if misplaced;
(b) the date of receipt;
(c) the activity at a specified date;
(d) the location of the substance, updated at appropriate intervals;
(e) the date and manner of transfer or disposal (when appropriate).

582 ACOP paragraph 578 sets out the frequency at which accountancy checks are normally required. Other examples of intervals at which the location of a source should be updated are:

(a) for static sources securely attached to machines, the interval between checks may be up to one month, providing that additional checks are carried out following any maintenance or repair which could have affected the source;
(b) for sources located within patients, the interval between checks should be compatible with the clinical treatment of that patient.

583 An annual check should be carried out to make sure the accounting record is correct. The check excludes any radioactive substances which have decayed to an insignificant quantity.

Accounting – specific situations

584 In production processes directly involving dispersible radioactive substances, the accounting procedures will vary with the scale of operation. The records held for production processing and waste disposal will normally be sufficient.

585 When accounting for higher-activity sources, it is acceptable to use unique seals where high dose rates may exist. Accountancy intervals should be subject to a suitable risk assessment and measures implemented to make sure that the process is as low as is reasonably practicable, while still maintaining a sufficient period of accounting.

Guidance 29(1)

586 In small-scale laboratories it should be sufficient to know the activity present and the radionuclides involved in each room, supported by the records required for waste disposal purposes.

587 Very small sources or articles with dimensions under 5 mm may be treated as dispersible radioactive substances for the purposes of accounting.

588 Accounting procedures will not be appropriate for radioactive substances where they:

(a)   have a half-life of less than three hours;
(b)   are in the form of contamination;
(c)   are in the form of a discrete source or confined radioactive substance, and their quantity or concentration does not exceed that specified in Parts 1 or 2 of Schedule 7, columns 2 and 3;
(d)   are dispersed in the body of a person, though this does not apply to samples taken from such people;
(e)   form part of a nuclear reactor, until such time as radioactive components are removed or the reactor is decommissioned; or
(f)   are undergoing irradiation in a nuclear reactor, particle accelerator or other similar device, provided they are not directly related to the operation of the plant.

Regulation 30 Keeping and moving of radioactive substances

Regulation 30(1)

(1)   An employer must ensure, so far as is reasonably practicable, that any radioactive substance under its control which is not for the time being in use or being moved, transported or disposed of –

(a)   is kept in a suitable receptacle; and
(b)   is kept in a suitable store.

Guidance 30(1)

589 In addition to this provision, employers registered under the Environmental Permitting (England and Wales) Regulations 2016[19] and the Radioactive Substances Act 1993[20] (Scotland) that do not have to register or license due to an exemption order must keep radioactive substances properly. In these circumstances, a single system of keeping can be used which satisfies both sets of legal requirements.

Suitable receptacle

590 If the radioactive substance is not in use, a receptacle for storage is suitable where it provides effective restriction of exposure, prevention of dispersal and physical security. A suitable receptacle should provide:

(a)   radiation shielding – it is advisable that the surface dose rate never exceeds 2 mSv per hour and usually it should be much less;
(b)   the ability to withstand damage from normal use and foreseeable misuse or accident;
(c)   fire resistance;
(d)   prevention of unauthorised access, exposure or dispersal.

591   Special considerations of receptacle design are required where:

(a)   the radioactive substance is corrosive, self-heating or pyrophoric;
(b)   there could be pressure build-up inside the receptacle;

(c)   the storage environment itself is corrosive.

Suitable store

592 A suitable store for radioactive substances should provide:

(a)   protection from the weather;
(b)   resistance to fire sufficient to minimise dispersal and loss of shielding, taking into account combustible materials in the vicinity and the likely temperatures that would be reached;
(c)   shielding to achieve the lowest dose rate that is reasonably practicable outside the store. Where non-classified persons may approach the outside of the store, it is advisable that the dose rate does not exceed 2.5 µSv per hour. It may need to be lower in special cases if employers wish to avoid designating the area as a supervised area (see regulation 17(3));
(d)   ventilation for both radioactive and non-radioactive substances that have accumulated as both may be harmful. Ventilation should also be provided for a radioactive substance that has been spilt or accidentally dispersed;
(e)   physical security so that access is only possible to people permitted by the employer.

593 Any store allocated to radioactive substances should only be used for such substances, their immediate containers and receptacles, and items such as handling tools and shielding material. Do not keep anything explosive or highly flammable in the store.

594 A sign prominently displayed on the outside of the store (preferably on the door) will warn people that the store may contain radioactive substances. Such signs must conform to the Health and Safety (Safety Signs and Signals) Regulations 1996.[8]

(2)   *An employer who causes or permits a radioactive substance to be moved (otherwise than by transporting it) must ensure that, so far as is reasonably practicable, the substance is kept in a suitable receptacle, suitably labelled, while it is being moved.*

Suitable receptacles for moving radioactive substances

595 This requirement particularly applies during site movements of radioactive material.

596 Transport is defined in regulation 2 and covers all conveyance through public places. Standards for packaging and labelling during transport can be found in the Carriage of Dangerous Goods and Use of Transportable Pressure Equipment Regulations 2009.[26]

597 If a site is open to the public, such as a hospital or university, movements of radioactive substances by vehicle within the site are defined as transport.

598 The following should be considered when assessing the suitability of a receptacle for movement:

(a)   the adequacy of the shielding to protect the person moving the substances;
(b)   the distance of the movement;
(c)   possible hazards encountered and the consequences of an incident, for example from spillage or dispersal;
(d)   the physical and chemical form and activity of the substances being moved.

599 Labelling must provide sufficient information for the safety of the person moving the receptacle and indicate the nature and activity of the substances being moved, or allow easy identification. The extent of the information will depend on circumstances, but taking appropriate action in the event of accidental spillage or dispersal would be information that should be made clear.

Regulation 30(3)

(3) Nothing in paragraphs (1) or (2) applies in relation to a radioactive substance while it is in or on the live body or corpse of a human being.

Regulation 31 Notification of certain occurrences

Regulation 31(1)–(2)

(1) An employer must immediately notify the appropriate authority in any case where a quantity of a radioactive substance which was under its control and which exceeds the quantity specified for that substance in column 5 of Part 1 of Schedule 7 –

(a) has been released or is likely to have been released into the atmosphere as a gas, aerosol or dust; or
(b) has been spilled or otherwise released in such a manner as to give rise to significant contamination.

(2) Paragraph (1) does not apply where such release –

(a) in relation to England and Wales –
  (i) was in accordance with an environmental permit under the Environmental Permitting (England and Wales) Regulations 2016 in respect of mobile radioactive apparatus within the meaning of those regulations;
  (ii) was in a manner specified in such an environmental permit in respect of radioactive waste within the meaning of those regulations; or
  (iii) did not, under regulation 12 of those regulations, require an environmental permit.
(b) in relation to Scotland –
  (i) was in accordance with a registration under section 10 of the Radioactive Substances Act 1993 or which was exempt from such registration by virtue of section 11 of that Act; or
  (ii) was in a manner specified in an authorisation to dispose of radioactive waste under section 13 of that Act or which was exempt from such authorisation by virtue of section 15 of that Act.

Guidance 31(1)–(2)

Notifying accidental releases and spillage

600 The term 'release' includes accidental spillages of radioactive substances, such as bench spills in a laboratory, and the term 'atmosphere' covers the internal environment of buildings as well as the external atmosphere. In general, where a dusty solid is released, it can be assumed for the purposes of regulation 31(1)(a) that the amount of dust released into the atmosphere is one-thousandth of the total amount involved in the occurrence. However, this assumption cannot be made where an exceptionally fine dust is involved or there are other reasons to judge it inappropriate, for instance the release has occurred under pressure or as a result of an explosion.

601 As a general rule, contamination arising from a spillage which exceeds the quantity specified in Schedule 7, column 5, is significant. The exception to this is where the spillage is in an enclosure or other such localised facility, designed, maintained and used to effectively prevent the release going beyond that facility. This exception applies, for instance, to glove boxes, purpose-designed enclosures

Guidance   31(1)–(2)

and benches in laboratories, and specially designed toilets in nuclear medicine departments. However, this exception does not cover releases affecting whole rooms or buildings where people work and could receive a significant exposure to ionising radiation as a result of the spillage.

Notifying other accidents etc

602 Employers must notify HSE about certain incidents involving equipment used in connection with medical exposures. In addition the Reporting of Injuries, Diseases and Dangerous Occurrences Regulations 2013 (RIDDOR)[27] require certain events to be reported to the relevant enforcing authority (usually HSE), particularly the malfunction of:

(a) a 'radiation generator' or its ancillary equipment used in fixed or mobile industrial radiography, the irradiation of food or the processing of products by irradiation, that causes it to fail to de-energise at the end of the intended exposure period ('radiation generator' means any electrical equipment emitting ionising radiation and containing components operating at a potential difference of more than 5 kV); or

(b) equipment used in fixed or mobile industrial radiography or gamma irradiation that causes a radioactive source to fail to return to its safe position by normal means at the end of the intended exposure period.

603 Even incidents listed above, where no one is exposed to ionising radiation, must be reported to HSE as a dangerous occurrence. Where a report is required under IRR17, there is no requirement to also report the incident under RIDDOR.[27]

Regulation   31(3)

(3)   Where an employer has reasonable cause to believe that a quantity of a radioactive substance which exceeds the quantity for that substance specified in column 6 of Part 1 of Schedule 7 and which was under its control is lost or has been stolen, the employer must immediately notify the appropriate authority of that loss or theft, as the case may be.

Guidance   31(3)

604 Employers with substances registered under EPR in England and Wales and RSA in Scotland, who keep radioactive material or operate under an EPR/RSA exemption order, must also report a loss or theft to the relevant Environment Agency.

605 Where accounting estimates involve large quantities of radioactive substances, as for example in the nuclear industry, uncertainties in those estimates can be of the order of the quantities specified in Part 1 of Schedule 7, column 6, thus apparently requiring notification to the appropriate authority under the terms of regulation 31(3). In such cases an agreement may be reached with the appropriate authority at local level as to what accountancy error is sufficiently large to constitute reasonable grounds for believing that the radioactive substances have been lost.

Regulation 31(4)–(5)

(4)   Where an employer suspects or has been informed that an occurrence notifiable under this regulation may have occurred, it must make an immediate investigation and, unless that investigation shows that no such occurrence has occurred, it must immediately make a notification under the relevant paragraph of this regulation.

(5)   An employer who makes any investigation in accordance with paragraph (4) must make a report of that investigation and must, unless the investigation showed that no such occurrence occurred, keep that report or a copy of the report for at least 30 years from the date on which it was made or, in any other case, for at least 2 years from the date on which it was made.

Regulation	31(6)

(6) In this regulation "appropriate authority" means –

(a) in relation to an occurrence notifiable under this regulation as a result of work carried out on nuclear premises, the ONR;

(b) otherwise, the Executive.

Regulation 32 Duties of manufacturers etc of articles for use in work with ionising radiation

Regulation	32(1)

(1) In the case of articles for use at work, where that work is work with ionising radiation, section 6(1) of the 1974 Act (which imposes general duties on manufacturers etc. as regards articles and substances for use at work) is modified so that any duty imposed on any person by that subsection includes a duty to ensure that any such article is so designed and constructed as to restrict so far as is reasonably practicable the extent to which employees and other persons are or are likely to be exposed to ionising radiation.

Guidance	32(1)

Application to medical exposures

606 Regulation 32(1) does not apply to the protection of those undergoing a medical examination or treatment. However, it does apply to the exposure of:

(a) employees who carry out those exposures;

(b) patients not subject to that medical exposure;

(c) members of the public.

Duties of manufacturers and suppliers to restrict exposure

607 Suppliers and manufacturers of any article containing a radioactive substance, including a sealed source, must make sure that suitable leak tests are carried out as soon as practicable and before it is supplied to the user.

608 The Health and Safety at Work Act 1974 (HSWA) defines an article as something that is designed for use by a person at work. It could include plant, machinery, equipment, appliance and any component of these.

609 The extension of the duty on manufacturers and suppliers under section 6(1) of the HSWA regarding articles for use at work does not apply to equipment used for medical exposures. This equipment will be subject to the requirements of the Medical Devices Regulations 2002 (as amended),[28] which are enforced by the Medicines and Healthcare Products Regulatory Agency.

610 Designers, manufacturers and suppliers must make sure that clear and complete information is passed along the supply chain, so that users are fully aware of how an article should be used or installed. This will allow the article to be operated as intended, to maintain or improve standards for restricting exposures (and achieving compliance with these Regulations).

611 Sometimes, a commercially available article that was neither designed for work with ionising radiation nor sold by the supplier for that purpose is used as such by an employer. One example would be an extractor fan, bought as a fan but then incorporated into a ventilation system intended to control airborne radioactive substances. Unless an article has been supplied on the basis of its future use or its design criteria, employers must make sure that it complies with the regulations and achieves the necessary performance standard.

Regulation	32(2)

(2)   Where a person erects or installs an article for use at work, being work with ionising radiation, that person must –

(a)   undertake a critical examination of the way in which the article was erected or installed for the purpose of ensuring, in particular, that –

(i)   any safety features and warning devices operate correctly; and

(ii)   there is sufficient protection for persons from exposure to ionising radiation;

(b)   consult with the radiation protection adviser that they appointed, or that the employer engaged in work with ionising radiation appointed, with regard to the nature and extent of any critical examination and the results of that examination; and

(c)   provide the employer engaged in work with ionising radiation with adequate information about proper use, testing and maintenance of the article.

Guidance	32(2)

Critical examination by installer/erector

612 A critical examination must be carried out if there are radiation protection implications arising from the way in which an article is being, or has been, erected or installed.

613 The requirement to carry out a critical examination also applies to second-hand articles, in addition to those that are new. The employer who erects or installs the second-hand article is responsible for carrying out the critical examination.

614 The critical examination may be carried out following erection or installation, during commissioning, or as part of trials prior to normal use. This may require co-operation between the various employers involved at each stage of a complex installation (see regulation 16). The employer who erects or installs the article is responsible for ensuring that the critical examination is carried out and not the user. This is still the case even if the critical examination is carried out, by agreement, during final trials under the supervision of the user's own RPA. Unlike regulation 32(1), the requirement to carry out a critical examination requires the dutyholder to consider the protection provided for people undergoing medical exposures, as well as the adequacy of protection for staff and members of the public.

615 Although the erector or installer must consult an RPA, regulation 32(2) does not require the RPA to be present when the critical examination is carried out, unless necessary for the conduct of the examination. This may be the installer's own RPA or one appointed by the employer who has purchased the equipment.

616 A critical examination must be carried out for articles containing radioactive substances and for X-ray generators. Other articles which may form part of the plant must also be covered by the critical examination. This should take into account shielding, ease of decontamination of surfaces, containment and any other aspects of radiation protection. However, the critical examination does not need to include matters which are not likely to be affected by the way the article has been erected or installed, such as the basic safety of the article.

Regulation	33

On 6 February 2018, regulation 33 (equipment used for medical exposure) was deleted from IRR17. Provisions relating to medical equipment are now contained in the Ionising Radiation (Medical Exposure) Regulations 2017 (see www.legislation.gov.uk).

Regulation 34 Misuse of or interference with sources of ionising radiation

Regulation	34

No person may intentionally or recklessly misuse or without reasonable excuse interfere with any radioactive substance or any electrical equipment to which these Regulations apply.

PART 7 Duties of employees and miscellaneous

Regulation 35 Duties of employees

Regulation 35

(1)   An employee who is engaged in work with ionising radiation must not knowingly expose themselves or any other person to ionising radiation to an extent greater than is reasonably necessary for the purposes of their work, and must exercise reasonable care while carrying out such work.

(2)   Every employee or outside worker for whom personal protective equipment is provided pursuant to regulation 9(2)(c) must –

(a)   make full and proper use of any such personal protective equipment;
(b)   immediately report to the employer who provided any such personal protective equipment any defect they discover in that equipment; and
(c)   take all reasonable steps to ensure that any such personal protective equipment is returned after use to the accommodation provided for it.

(3)   It is the duty of every classified outside worker not to misuse the radiation passbook issued to that worker or falsify or attempt to falsify any of the information contained in it.

(4)   Any employee to whom regulation 22(1) or regulation 13(2)(b) relates must comply with any reasonable requirement imposed on that person by that person's employer for the purposes of making the measurements and assessments required under regulation 22(1) and regulation 24(1).

(5)   An employee who is subject to medical surveillance under regulation 25 must, when required by their employer and at the cost of the employer, present themselves during their working hours for such medical examination and tests as may be required for the purposes of regulation 25 (2) and must provide the relevant doctor with such information concerning their health as the relevant doctor may reasonably require.

(6)   Where an employee has reasonable cause to believe that –

(a)   they or some other person has received an overexposure; or
(b)   an occurrence mentioned in paragraph (1) or (3) of regulation 31 has occurred;
they must immediately notify their employer of that belief.

Guidance 35

617   This regulation places duties on employees and outside workers when they are engaged in work with ionising radiation. Workers must take reasonable care to make sure that they are working safely. Employers are responsible under regulation 15 to make sure employees and outside workers understand their legal duties with regard to their health and safety.

Application to medical exposures

618 This regulation does not apply to the protection of people undergoing a medical examination or treatment. However, it does apply to employees who carry out those exposures.

Other duties of employees

619 Employees must:

(a)   correctly use the PPE provided by their employer, even for jobs that will only take a couple of minutes;
(b)   take care of PPE and store it correctly;
(c)   tell their employer about any faults with PPE and report any damage;
(d)   take care of their radiation passbook and not enter false information;
(e)   co-operate with employers about dose measurements and assessments;
(f)   report loss or damage of personal dosemeters to their employer immediately;
(g)   co-operate with their employer and doctor in completing medical surveillance;
(h)   tell their employer about actual or suspected incidents which the employer has a duty to investigate, such as apparent overexposures or loss of a source.

Co-operation

620 Co-operating means that in accordance with HSWA, employees must wear and take reasonable care of personal dosemeters that have been provided by their employer. This includes wearing them, and returning them for processing in accordance with their employer's instructions. An employee may be committing an offence under section 7 of HSWA if they fail to co-operate with their employer and wear and take care of their personal dosemeters and own health and safety.

Duties of employers

621 Employers can only meet certain specific duties under these Regulations with the co-operation of employees and outside workers. Employers must therefore make sure that employees and outside workers understand the risks associated with their work, and what they are required to do to control them. Employers should check that employees are complying with the required control measures and challenge and correct unsafe behaviour. See HSE's *Managing for health and safety* for more information: www.hse.gov.uk/pubns/books/hsg65.htm.

Regulation 36 Approval of dosimetry services

(1)   The Executive (or such other person as may from time to time be specified in writing by the Executive) may, by a certificate in writing, approve (in accordance with such criteria as may from time to time be specified by the Executive) a suitable dosimetry service for such of the purposes of these Regulations or of the Radiation (Emergency Preparedness and Public Information) Regulations 2001 as are specified in the certificate.

(2)   A certificate made pursuant to paragraph (1) may be made subject to conditions and may be revoked in writing at any time.

(3)   The Executive (or such other person as may from time to time be specified in writing by the Executive) may at such periods as it considers appropriate carry out a re-assessment of any approval granted pursuant to paragraph (1).

622 Further information can be found on HSE's website at: www.hse.gov.uk/radiation/ionising/dosimetry/ads.htm.

Regulation 37 Defence on contravention

Regulation 37(1)–(6)

(1)   In any proceedings against an employer for an offence under regulation 5(2) (notification), 6(3) (registration) or 7(2) (consent), it is a defence for that employer to prove that –

(a)   it neither knew nor had reasonable cause to believe that it had carried out or might be required to carry out work that required notification under regulation 5(2), registration under regulation 6(3) or consent under regulation 7(2) (as the case may be); and

(b)   in a case where it discovered that it had carried out or was carrying out such work, it had immediately notified, registered or applied for consent for such work (as the case may be) in accordance with those regulations.

(2)   The defence in paragraph (1) –

(a)   in connection with an offence under regulation 6(3), does not apply in relation to the operation of a radiation generator; and

(b)   in connection with an offence under regulation 7(2), only applies in relation to a practice referred to in regulation 7(1)(g).

(3)   In any proceedings against an employer for an offence under regulation 8, it is a defence for that employer to prove that –

(a)   it neither knew nor had reasonable cause to believe that it had commenced a new activity involving work with ionising radiation; and

(b)   in a case where it had discovered that it had commenced a new activity involving work with ionising radiation, it had as soon as practicable made an assessment as required by regulation 8.

(4)   In any proceedings against an employer for an offence under regulation 28(2) it is a defence for that employer to prove that –

(a)   it had received and reasonably relied on a written undertaking from the supplier of the article concerned that the article complied with the requirements of that paragraph; and

(b)   it had complied with the requirements of paragraph (3) of that regulation.

(5)   In any proceedings against an employer of an outside worker for a breach of a duty under these Regulations it is a defence for that employer to show that –

(a)   it had entered into a contract in writing with the employer who had designated an area as a controlled or supervised area and in which the outside worker was working or was to work for that employer to perform that duty on their behalf; and

(b)   the breach of duty was a result of the failure of the employer referred to in sub-paragraph (a) to fulfil that contract.

(6)   In any proceedings against any employer who has designated a controlled or supervised area in which any outside worker is working or is to work for a breach of a duty under these Regulations it is a defence for that employer to show that –

(a)   it had entered into a contract in writing with the employer of an outside worker for that employer to perform that duty on its behalf; and

Regulation 37(6)–(9)

(b)   the breach of duty was a result of the failure of the employer referred to in sub-paragraph (a) to fulfil that contract.

(7)   A person charged is not, without the permission of the court, entitled to rely on the defence referred to in paragraph (5) or (6) unless, within a period ending seven clear days before the hearing, that person has served on the prosecutor a notice in writing of that person's intention to rely on the defence and the notice must be accompanied by a copy of the contract on which that person intends to rely and, if that contract is not in English, an accurate translation of that contract into English.

(8)   Where a contravention of these Regulations by any person is due to the act or default of some other person, that other person will be guilty of the offence which would, but for any defence under this regulation available to the first-mentioned person, be constituted by the act or default.

(9)   In this regulation, "appropriate authority" means –

(a)   in connection with the application of this regulation in relation to, or in relation to any activity carried out on, any nuclear premises, the ONR;

(b)   otherwise, the Executive.

Regulation 38 Exemption certificates

Regulation 38(1)–(3)

(1)   Subject to paragraph (2), the appropriate authority may, by a certificate in writing, exempt –

(a)   any person or class of persons;

(b)   any premises or class of premises; or

(c)   any equipment, apparatus or substance or class of equipment, apparatus or substance,

from any requirement or prohibition imposed by these Regulations and any such exemption may be granted subject to conditions and to a limit of time and may be revoked by a certificate in writing at any time.

(2)   The appropriate authority must not grant an exemption unless, having regard to the circumstances of the case and in particular to –

(a)   the conditions, if any, which it proposes to attach to the exemption; and

(b)   any other requirements imposed by or under any enactments which apply to the case,

it is satisfied that –

(c)   the health and safety of persons who are likely to be affected by the exemption will not be prejudiced in consequence of it; and

(d)   compliance with the fundamental radiation protection provisions underlying regulations 9(1) and (2)(a), 12, 13(1), 17(1) and (3), 20(1), 21(1), 22(1) and 25(2) will be achieved.

(3)   In this regulation, "appropriate authority" means –

(a)   in connection with the application of this regulation in relation to, or in relation to any activity carried out on, any nuclear premises, the ONR;

(b)   otherwise, the Executive.

Regulation 39 Extension outside Great Britain

(1)   Subject to paragraph (2), these Regulations apply to any work outside Great Britain to which sections 1 to 59 and 80 to 82 of the 1974 Act apply by virtue of the Health and Safety at Work etc. Act 1974 (Application outside Great Britain) Order 2013 as they apply to work within Great Britain.

(2)   For the purposes of paragraph (1), in any case where it is not reasonably practicable for an employer to comply with the requirements of these Regulations in so far as they relate to functions being performed by a relevant doctor or by an approved dosimetry service, it is sufficient compliance with any such requirements if the employer makes arrangements affording an equivalent standard of protection for its employees and those arrangements are set out in local rules.

Regulation 40 Modifications relating to the Ministry of Defence etc

(1)   In this regulation, any reference to –

(a)   "visiting forces" is a reference to visiting forces within the meaning of any provision of Part 1 of the Visiting Forces Act 1952; and

(b)   "headquarters or organisation" is a reference to a headquarters or organisation designated for the purposes of the International Headquarters and Defence Organisations Act 1964.

(2)   The Secretary of State for Defence may, in the interests of national security, by a certificate in writing exempt –

(a)   Her Majesty's Forces;

(b)   visiting forces;

(c)   any member of a visiting force working in or attached to any headquarters or organisation; or

(d)   any person engaged in work with ionising radiation for, or on behalf of, the Secretary of State for Defence,

from all or any of the requirements or prohibitions imposed by these Regulations and any such exemption may be granted subject to conditions and to a limit of time and may be revoked at any time by a certificate in writing, except that, where any such exemption is granted, suitable arrangements must be made for the assessment and recording of doses of ionising radiation received by persons to whom the exemption relates.

(3)   Regulations 5, 6 and 7 do not apply in relation to work carried out by visiting forces or any headquarters or organisation on premises under the control of such visiting force, headquarters or organisation, as the case may be, or on premises under the control of the Secretary of State for Defence.

(4)   With respect to any work with ionising radiation undertaken for, or on behalf of the Secretary of State for Defence –

(a)   the requirement of regulation 5(2) and (3), 6(4)(a) and (b), and 7(3)(a) and (b) to notify particulars specified by the appropriate authority (as defined for the purposes of those regulations) only applies in relation to the particulars that may be so specified from the list set out in paragraph (9); and

Regulation 40

(b) any requirement to provide the particulars described in paragraphs (9) (d), (e), (f), (g), (h), (i), and (k) does not apply where –

    (i) the Secretary of State for Defence decides that the provision of such particulars will be contrary to the interests of national security; or

    (ii) suitable alternative arrangements have been agreed with the appropriate authority (as defined in paragraph (10)).

(5) Regulation 5(4) does not apply to an employer in relation to work with ionising radiation undertaken for or on behalf of the Secretary of State for Defence, visiting forces or any headquarters or organisation.

(6) Sub-paragraph (i) of regulation 22(3) does not apply in relation to a practice carried out –

(a) by or on behalf of the Secretary of State for Defence;

(b) by a visiting force; or

(c) by any member of a visiting force in or attached to any headquarters or organisation.

(7) Regulations 23(6), (7) and (8) and regulation 25(8) do not apply in relation to visiting forces or any member of a visiting force working in or attached to any headquarters or organisation.

(8) In regulation 26(1) the requirement to notify the relevant authority (as defined for the purposes of that regulation) of a suspected overexposure and the results of the consequent investigation and assessment do not apply in relation to the exposure of –

(a) a member of a visiting force; or

(b) a member of a visiting force working in or attached to a headquarters or organisation.

(9) The particulars referred to in paragraph (4) are –

(a) the name, address, telephone number and e-mail address of the employer;

(b) the address of the premises where or from where the work activity is to be carried out and a telephone number or e-mail address for such premises;

(c) the nature of the business of the employer;

(d) a description of the work with ionising radiation;

(e) particulars of the source or sources of ionising radiation including the type of electrical equipment used or operated and the nature of any radioactive substance;

(f) the quantities of any radioactive substance used in the work;

(g) the identity of any person engaged in the work;

(h) whether or not any source is to be used at premises other than the address given in sub-paragraph (b);

(i) the location and description of any premises at which the work is carried out on each occasion that it is so carried out;

(j) the date of notification, registration or application for consent to carry out the work activity and the date of commencement of the work activity;

(k) the duration of any period over which the work is carried out and the date of termination of the work activity.

Regulation	40

(10) In paragraph (4)(b)(ii), "appropriate authority" means –

(a) In connection with the application of this regulation to, or in relation to any activity carried out on, any nuclear premises, the ONR;

(b) Otherwise, the Executive.

Regulation 41 Transitional provisions and savings

Regulation	41

Schedule 8, which makes transitional provisions and savings, has effect.

Regulation 42 Modifications and revocation

Regulation	42

(1) Schedule 9, which contains modifications to primary legislation and instruments, has effect.

(2) The Ionising Radiations Regulations 1999 are revoked.

Regulation 43 Review

Regulation	43

(1) The Secretary of State must from time to time –

(a) carry out a review of the regulatory provision contained in these Regulations, and

(b) publish a report setting out the conclusions of the review.

(2) The first report must be published before 1st January 2023.

(3) Subsequent reports must be published at intervals not exceeding 5 years.

(4) Section 30(3) of the Small Business, Enterprise and Employment Act 2015 requires that a review carried out under this regulation must, so far as is reasonable, have regard to how the Directive is implemented in other member States.

(5) Section 30(4) of the Small Business, Enterprise and Employment Act 2015 requires that a report published under this regulation must, in particular –

(a) set out the objectives intended to be achieved by the regulatory provision referred to in paragraph (1)(a);

(b) assess the extent to which those objectives are achieved;

(c) assess whether those objectives remain appropriate; and

(d) if those objectives remain appropriate, assess the extent to which they could be achieved in another way which involves less onerous regulatory provision.

(6) In this regulation, "regulatory provision" has the same meaning as in sections 28 to 32 of the Small Business, Enterprise and Employment Act 2015 (see section 32 of that Act).

Schedule 1 Work not required to be notified under regulation 5

Schedule	1

Regulations 5(1), 6(2) and 14(3)

(1)   Work with ionising radiation is not required to be notified in accordance with regulation 5 when the only such work being carried out is in one or more of the following categories –

(a)   *where the concentration of activity per unit mass of a radioactive substance does not exceed the concentration specified in column 2 of Part 1 of Schedule 7 (for artificial radionuclides and naturally occurring radionuclides which are processed for their radioactive, fissile or fertile properties) or column 2 of Part 2 of Schedule 7 (for naturally occurring radionuclides which are not processed for their radioactive, fissile or fertile properties);*

(b)   *where the quantity of radioactive substance involved does not exceed the quantity specified in column 3 of Part 1 of Schedule 7 (for artificial radionuclides and naturally occurring radionuclides which are processed for their radioactive, fissile or fertile properties) or column 3 of Part 2 of Schedule 7 (for naturally occurring radionuclides which are not processed for their radioactive, fissile or fertile properties);*

(c)   *where the concentration of activity per unit mass or quantity of a radioactive substance does not exceed values which may be approved by the appropriate authority for specific types of work and where such work satisfies the exemption criteria set out in paragraphs 2 and 3 below;*

(d)   *where apparatus contains radioactive substances in a quantity exceeding the values specified in sub-paragraphs (a) and (b) provided that –*

(i)   *the apparatus is of a type approved by the Executive;*

(ii)   *the apparatus is constructed in the form of a sealed source;*

(iii)   *the apparatus does not under normal operating conditions cause a dose rate of more than 1 μSvh^{-1} at a distance of 0.1 m from any accessible surface; and*

(iv)   *conditions for the disposal of the apparatus have been specified by the relevant environmental body;*

(e)   *the operation of any electrical apparatus to which these Regulations apply other than apparatus referred to in sub-paragraph (f) provided that –*

(i)   *the apparatus is of a type approved by the Executive; and*

(ii)   *the apparatus does not under normal operating conditions cause a dose rate of more than 1 μSvh^{-1} at a distance of 0.1 m from any accessible surface;*

(f)   *the operation of –*

(i)   *any cathode ray tube intended for the display of visual images; or*

(ii)   *any other electrical apparatus operating at a potential difference not exceeding 30kV,*

provided that the operation of the tube or apparatus does not under normal operating conditions cause a dose rate of more than 1 μSvh^{-1} at a distance of 0.1 m from any accessible surface; or

Schedule   **1**

(g)   where the work involves contaminated material resulting from authorised releases which the relevant environmental body has declared not to be subject to further control.

(2)   The criteria for the exemption from notification of work with ionising radiation are as follows:

(a)   the radiological risks to individuals caused by such work are sufficiently low as to be of no regulatory concern;

(b)   work of such type has been found to be justified; and

(c)   such work is inherently safe.

(3)   Work with ionising radiation only meets the requirements of paragraph 2(a) if –

(a)   in relation to an employee, the effective dose caused by such work does not exceed 1 mSv in a calendar year; and

(b)   in relation to any other person, the following requirements are met in all circumstances where it is reasonably practicable to do so –

(i)   the effective dose caused by such work from radionuclides which are not naturally occurring radionuclides does not exceed 10 µSv in a calendar year; and

(ii)   the effective dose caused by such work from naturally occurring radionuclides does not exceed 1 mSv in a calendar year.

(4)   In paragraph 2(b), "found to be justified" has the meaning given by regulation 4(4) of the Justification of Practices Involving Ionising Radiation Regulations 2004.

(5)   In this Schedule –

"appropriate authority" means –

(a)   in relation to any activity carried out exclusively or primarily on premises which are or are on –
(i)   an authorised defence site;
(ii)   a new nuclear build site;
(iii)   a nuclear warship site,
the ONR;

(b)   otherwise, the Executive;

"relevant environmental body" –

(a)   in relation to England, means the Environment Agency;
(b)   in relation to Wales, means the Natural Resources Body for Wales;
(c)   in relation to Scotland, means the Scottish Environment Protection Agency.

Schedule 2 Consent to carry out a practice: indicative list of information

Schedule	2

Regulation 7(3)

(1) Responsibilities and organisational arrangements for protection and safety.

(2) Staff competences, including information and training.

(3) Design features of the facility and of radiation sources.

(4) Anticipated occupational and public exposures in normal operation.

(5) Safety assessment of the activities and the facility in order to –

    (a) identify ways in which potential exposures or accidental and unintended medical exposures could occur; estimate, to the extent practicable, the probabilities and magnitude of potential exposures;
    (b) estimate, to the extent practicable, the probabilities and magnitude of potential exposures;
    (c) assess the quality and extent of protection and safety provisions, including engineering features, as well as administrative procedures;
    (d) define the operational limits and conditions of operation.

(6) Emergency procedures.

(7) Maintenance, testing, inspection and servicing so as to ensure that the radiation source and the facility continue to meet the design requirements, operational limits and conditions of operation throughout their lifetime.

(8) Management of radioactive waste and arrangements for the disposal of such waste, in accordance with applicable regulatory requirements.

(9) Management of disused sources.

(10) Quality assurance.

Schedule 3 Dose limits

Schedule	3

Regulations 2(1) and 12

Part 1 – Classes of persons to whom dose limits apply

Employees and trainees of 18 years of age or above

(1)   For the purposes of regulation 12(1), the limit on effective dose for any employee or trainee, being of 18 years of age or above, is 20 mSv in any calendar year.

(2)   Without prejudice to paragraph 1 –

    (a)   the limit on equivalent dose for the lens of the eye is –
        (i)   20 mSv in a calendar year; or
        (ii)   in accordance with conditions approved by the Executive from time to time, 100 mSv in any period of five consecutive calendar years subject to a maximum equivalent dose of 50 mSv in any single calendar year;
    (b)   the limit on equivalent dose for the skin is 500 mSv in a calendar year as applied to the dose averaged over any area of 1 cm^2 regardless of the area exposed;
    (c)   the limit on equivalent dose for the extremities is 500 mSv in a calendar year.

Trainees aged under 18 years

(3)   For the purposes of regulation 12(1), the limit on effective dose for any trainee under 18 years of age is 6 mSv in any calendar year.

(4)   Without prejudice to paragraph 3 –

    (a)   the limit on equivalent dose for the lens of the eye is 15 mSv in a calendar year;
    (b)   the limit on equivalent dose for the skin is 150 mSv in a calendar year as applied to the dose averaged over any area of 1 cm^2 regardless of the area exposed;
    (c)   the limit on equivalent dose for the extremities is 150 mSv in a calendar year.

Other persons

(5)   Subject to paragraph 6, for the purposes of regulation 12(1) the limit on effective dose for any person other than an employee or trainee referred to in paragraph 1 or 3, including any person below the age of 16, is 1 mSv in any calendar year.

(6)   Paragraph 5 does not apply in relation to any person (not being a carer and comforter) who may be exposed to ionising radiation resulting from the medical exposure of another and in such a case the limit on effective dose for any such person is 5 mSv in any period of 5 consecutive calendar years.

(7)   Without prejudice to paragraphs 5 and 6 –

Schedule 3

(a)   the limit on equivalent dose for the lens of the eye is 15 mSv in any calendar year;

(b)   the limit on equivalent dose for the skin is 50 mSv in any calendar year averaged over any 1 cm² area regardless of the area exposed;

(c)   the limit on equivalent dose for the extremities is 50 mSv in a calendar year.

Part 2

(8)   For the purposes of regulation 12(2), the limit on effective dose for employees or trainees of 18 years or above is 100 mSv in any period of five consecutive calendar years subject to a maximum effective dose of 50 mSv in any single calendar year.

(9)   Without prejudice to paragraph 8 –

(a)   the limit on equivalent dose for the lens of the eye is –
(i)   20 mSv in a calendar year; or
(ii) in accordance with conditions approved by the Executive from time to time, 100 mSv in any period of five consecutive calendar years subject to a maximum equivalent dose of 50 mSv in any single calendar year;

(b)   the limit on equivalent dose for the skin is 500 mSv in a calendar year as applied to the dose averaged over any area of 1 cm² regardless of the area exposed;

(c)   the limit on equivalent dose for the extremities is 500 mSv in a calendar year.

(10)   The employer must ensure that any employee in respect of whom regulation 12(2) applies is not exposed to ionising radiation to an extent that any dose limit specified in paragraphs 8 or 9 is exceeded.

(11)   An employer must not put into effect a system of dose limitation pursuant to regulation 12(2) unless –

(a)   the radiation protection adviser and any employees who are affected have been consulted;

(b)   any employees affected and the approved dosimetry service have been informed in writing of the decision and of the reasons for that decision; and

(c)   notice has been given to the appropriate authority at least 28 days (or such shorter period as the appropriate authority may allow) before the decision is put into effect giving the reasons for the decision.

(12)   Where there is reasonable cause to believe that any employee has been exposed to an effective dose greater than 20 mSv in any calendar year, the employer must, as soon as is practicable –

(a)   undertake an investigation into the circumstances of the exposure for the purpose of determining whether the dose limit referred to in paragraph 8 is likely to be complied with; and

(b)   notify the appropriate authority of that suspected exposure.

(13)   An employer must review the decision to put into effect a system of dose limitation pursuant to regulation 12(2) at appropriate intervals and in any event not less than once every five years.

Schedule 3

(14) Where as a result of a review undertaken pursuant to paragraph 13 an employer proposes to revert to a system of annual dose limitation pursuant to regulation 12(1), the provisions of paragraph 11 apply as if the reference in that paragraph to regulation 12(2) was a reference to regulation 12(1).

(15) Where an employer puts into effect a system of dose limitation in pursuance of regulation 12(2), the employer must record the reasons for that decision and must ensure that the record is preserved until any person subject to the system of dose limitation under regulation 12(2) has or would have attained the age of 75 years but in any event for at least 30 years from the making of the record.

(16) In any case where –

(a) the dose limits specified in paragraph 8 are being applied by an employer in respect of an employee; and

(b) the appropriate authority is not satisfied that it is impracticable for that employee to be subject to the dose limit specified in paragraph 1 of Part 1 of this Schedule,

the appropriate authority may require the employer to apply the dose limit specified in paragraph 1 of Part 1 with effect from such time as the appropriate authority may consider appropriate having regard to the interests of the employee concerned.

(17) In any case where, as a result of a review undertaken pursuant to paragraph 13, an employer proposes to revert to an annual dose limitation in accordance with regulation 12(1), the appropriate authority may require the employer to defer the implementation of that decision to such time as the appropriate authority may consider appropriate having regard to the interests of the employee concerned.

(18) Any person who is aggrieved by the decision of the appropriate authority taken pursuant to paragraphs 16 or 17 may appeal to the Secretary of State.

(19) Sub-sections (2) to (6) of section 44 of the 1974 Act apply for the purposes of paragraph 18 as they apply to an appeal under section 44(1) of that Act.

(20) The Health and Safety Licensing Appeals (Hearings Procedure) Rules 1974), as respects England and Wales, and the Health and Safety Licensing Appeals (Hearings Procedure) (Scotland) Rules 1974, as respects Scotland, apply to an appeal under paragraph 18 as they apply to an appeal under section 44(1) of the 1974 Act, but with the modification that references to a licensing authority are to be read as references to the appropriate authority.

(21) In this Part, "appropriate authority" means –

(a) in connection with the application of this Part in relation to, or in relation to any activity carried out on, any nuclear premises, the ONR;

(b) otherwise, the Executive.

Schedule 4 Matters in respect of which a radiation protection adviser must be consulted

Schedule	4

Regulation 14(1)

(1)   The implementation of requirements as to controlled and supervised areas.

(2)   The prior examination of plans for installations and the acceptance into service of new or modified sources of ionising radiation in relation to any engineering controls, design features, safety features and warning devices provided to restrict exposure to ionising radiation.

(3)   The regular calibration of equipment provided for monitoring levels of ionising radiation and the regular checking that such equipment is serviceable and correctly used.

(4)   The periodic examination and testing of engineering controls, design features, safety features and warning devices and regular checking of systems of work provided to restrict exposure to ionising radiation.

Schedule 5 Particulars to be entered in the radiation passbook

Schedule 5	**Regulation 22(5)**

(1) Individual serial number of the passbook.

(2) A statement that the passbook has been approved by the Executive for the purpose of these Regulations.

(3) Date of issue of the passbook by the approved dosimetry service.

(4) The name, telephone number and mark of endorsement of the issuing approved dosimetry service.

(5) The name, address, telephone number and e-mail address of the employer.

(6) Full name (surname, forenames), date of birth, gender and national insurance number of the classified outside worker to whom the passbook has been issued.

(7) Date of the last medical review of the classified outside worker and the relevant classification in the health record maintained under regulation 25 as fit, fit subject to conditions (which must be specified) or unfit.

(8) The relevant dose limits applicable to the classified outside worker to whom the passbook has been issued.

(9) The cumulative dose assessment in mSv for the year to date for the classified outside worker, external (whole body, organ or tissue) and/or internal as appropriate and the date of the end of the last assessment period.

(10) In respect of services performed by the classified outside worker –

> *(a) the name and address of the employer responsible for the controlled area;*
> *(b) the period covered by the performance of the services;*
> *(c) the following estimated dose information, as appropriate –*
>> *(i) an estimate of any whole body effective dose in mSv received by the classified outside worker;*
>> *(ii) in the event of non-uniform exposure, an estimate of the equivalent dose in mSv to organs and tissues as appropriate; and*
>> *(iii) in the event of internal contamination, an estimate of the activity taken in or the committed dose.*

Schedule 6 Particulars to be contained in a health record

Schedule	6

Regulation 25(2)(b)

The following particulars must be contained in a health record made for the purposes of regulation 25(3)

(a)   the employee's –
  (i)   full name;
  (ii)   sex;
  (iii)   date of birth;
  (iv)   permanent address; and
  (v)   National Insurance number;

(b)   the date of the employee's commencement as a classified person in present employment;

(c)   the nature of the employee's employment;

(d)   the date and type of the last medical examination or health review carried out in respect of the employee;

(e)   a statement by the relevant doctor made as a result of the last medical examination or health review carried out in respect of the employee classifying the employee as fit, fit subject to conditions (which should be specified) or unfit;

(f)   in relation to each medical examination and health review, the name and signature of the relevant doctor;

(g)   the name and address of the approved dosimetry service with whom arrangements have been made for maintaining the dose record in accordance with regulation 22.

Schedule 7 Quantities and concentrations of radionuclides

Part 1 Regulations 2(4), 6(2), 31(1), 31(3) and Schedule 1

Table of artificial radionuclides and naturally occurring radionuclides (which are processed for their radioactive, fissile or fertile properties)

1	2	3	4	5	6
Radionuclide name, symbol, isotope	Concentration for: Notification (any amount of radioactive material); Registration (amounts of radioactive material that exceed 1,000kg) Regulation 5(1) and Schedule 1, paragraph 1(a); regulation 6(2)(f) (Bq/g)	Quantity for Notification Regulation 5(1) and Schedule 1, paragraph 1(b) (Bq)	Concentration for Registration (amounts of radioactive material that do not exceed 1,000kg) Regulation 6(2)(e) (Bq/g)	Quantity for notification of occurrences Regulation 31(1) (Bq)	Quantity for notification of occurrences Regulation 31(3) (Bq)
Hydrogen					
H-3 (tritiated compounds)	10^2	10^9	10^6	10^{12}	10^{10}
Beryllium					
Be-7	10	10^7	10^3	10^{12}	10^8
Carbon					
C-11	0.01	10^6	10	10^{13}	10^7
C-11 (monoxide)	0.01	10^9	10	10^{12}	10^{10}
C-11 (dioxide)	0.01	10^9	10	10^{12}	10^{10}
C-14	1	10^7	10^4	10^{11}	10^8
Oxygen					
O-15	0.01	10^9	10^2	10^{10}	
Fluorine					
F-18	10	10^6	10	10^{13}	10^7
Sodium					
Na-22	0.1	10^6	10	10^{10}	10^7
Na-24	0.1	10^5	10	10^{11}	10^6
Silicon					
Si-31	10^3	10^6	10^3	10^{13}	10^7

Schedule 7

	1	2	3	4	5	6
	Radionuclide name, symbol, isotope	Concentration for: Notification (any amount of radioactive material); Registration (amounts of radioactive material that exceed 1,000kg) Regulation 5(1) and Schedule 1, paragraph 1(a); regulation 6(2)(f) (Bq/g)	Quantity for Notification Regulation 5(1) and Schedule 1, paragraph 1(b) (Bq)	Concentration for Registration (amounts of radioactive material that do not exceed 1,000kg) Regulation 6(2)(e) (Bq/g)	Quantity for notification of occurrences Regulation 31(1) (Bq)	Quantity for notification of occurrences Regulation 31(3) (Bq)
Phosphorus						
	P-32	10^3	10^5	10^3	10^{10}	10^6
	P-33	10^3	10^8	10^5	10^{11}	10^9
Sulphur						
	S-35	10^2	10^8	10^5	10^{11}	10^9
Chlorine						
	Cl-36	1	10^6	10^4	10^{10}	10^7
	Cl-38	10	10^5	10	10^{13}	10^6
Argon						
	Ar-37	0.01	10^8	10^6	10^{13}	
	Ar-41	0.01	10^9	10^2	10^9	
Potassium						
	K-40[(1)]	1	10^6	10^2	10^{10}	10^7
	K-42	10^2	10^6	10^2	10^{12}	10^7
	K-43	10	10^6	10	10^{11}	10^7
Calcium						
	Ca-45	10^2	10^7	10^4	10^{10}	10^8
	Ca-47	10	10^6	10	10^{11}	10^7
Scandium						
	Sc-46	0.1	10^6	10	10^{10}	10^7
	Sc-47	10^2	10^6	10^2	10^{11}	10^7
	Sc-48	1	10^5	10	10^{11}	10^6
Vanadium						
	V-48	1	10^5	10	10^{10}	10^6
Chromium						
	Cr-51	10^2	10^7	10^3	10^{12}	10^8
Manganese						
	Mn-51	10	10^5	10	10^{13}	10^6
	Mn-52	1	10^5	10	10^{10}	10^6
	Mn-52m	10	10^5	10	10^{13}	10^6
	Mn-53	10^2	10^9	10^4	10^{12}	10^{10}
	Mn-54	0.1	10^6	10	10^{11}	10^7
	Mn-56	10	10^5	10	10^{12}	10^6

Schedule 7

	1	2	3	4	5	6
	Radionuclide name, symbol, isotope	Concentration for: Notification (any amount of radioactive material); Registration (amounts of radioactive material that exceed 1,000kg) Regulation 5(1) and Schedule 1, paragraph 1(a); regulation 6(2)(f) (Bq/g)	Quantity for Notification Regulation 5(1) and Schedule 1, paragraph 1(b) (Bq)	Concentration for Registration (amounts of radioactive material that do not exceed 1,000kg) Regulation 6(2)(e) (Bq/g)	Quantity for notification of occurrences Regulation 31(1) (Bq)	Quantity for notification of occurrences Regulation 31(3) (Bq)
Iron						
Fe-52+	10	10^6	10	10^{12}	10^7
Fe-55	10^3	10^6	10^4	10^{11}	10^7
Fe-59	1	10^6	10	10^{10}	10^7
Cobalt						
Co-55	10	10^6	10	10^{11}	10^7
Co-56	0.1	10^5	10	10^{10}	10^6
Co-57	1	10^6	10^2	10^{11}	10^7
Co-58	1	10^6	10	10^{10}	10^7
Co-58m	10^4	10^7	10^4	10^{13}	10^8
Co-60	0.1	10^5	10	10^{10}	10^6
Co-60m	10^3	10^6	10^3	10^{16}	10^7
Co-61	10^2	10^6	10^2	10^{13}	10^7
Co-62m	10	10^5	10	10^{13}	10^6
Nickel						
Ni-59	10^2	10^8	10^4	10^{11}	10^9
Ni-63	10^2	10^8	10^5	10^{11}	10^9
Ni-65	10	10^6	10	10^{13}	10^7
Copper						
Cu-64	10^2	10^6	10^2	10^{12}	10^7
Zinc						
Zn-65	0.1	10^6	10	10^{10}	10^7
Zn-69	10^3	10^6	10^4	10^{14}	10^7
Zn-69m+	10	10^6	10^2	10^{12}	10^7
Gallium						
Ga-68	0.01	10^5	10	10^{13}	10^6
Ga-72	10	10^5	10	10^{11}	10^6
Germanium						
Ge-68+	0.01	10^5	10	10^{10}	10^6
Ge-71	10^4	10^8	10^4	10^{13}	10^9
Arsenic						
As-73	10^3	10^7	10^3	10^{11}	10^8
As-74	10	10^6	10	10^{11}	10^7
As-76	10	10^5	10^2	10^{11}	10^6

Schedule 7

1	2	3	4	5	6
Radionuclide name, symbol, isotope	Concentration for: Notification (any amount of radioactive material); Registration (amounts of radioactive material that exceed 1,000kg) Regulation 5(1) and Schedule 1, paragraph 1(a); regulation 6(2)(f) (Bq/g)	Quantity for Notification Regulation 5(1) and Schedule 1, paragraph 1(b) (Bq)	Concentration for Registration (amounts of radioactive material that do not exceed 1,000kg) Regulation 6(2)(e) (Bq/g)	Quantity for notification of occurrences Regulation 31(1) (Bq)	Quantity for notification of occurrences Regulation 31(3) (Bq)
As-77	10^3	10^6	10^3	10^{12}	10^7
Selenium					
Se-75	1	10^6	10^2	10^{11}	10^7
Bromine					
Br-82	1	10^6	10	10^{11}	10^7
Krypton					
Kr-74	0.01	10^9	10^2	10^9	
Kr-76	0.01	10^9	10^2	10^{10}	
Kr-77	0.01	10^9	10^2	10^9	
Kr-79	0.01	10^5	10^3	10^{10}	
Kr-81	0.01	10^7	10^4	10^{11}	
Kr-83m	0.01	10^{12}	10^5	10^{12}	
Kr-85	0.01	10^4	10^5	10^{12}	
Kr-85m	0.01	10^{10}	10^3	10^{10}	
Kr-87	0.01	10^9	10^2	10^9	
Kr-88	0.01	10^9	10^2	10^9	
Rubidium					
Rb-86	10^2	10^5	10^2	10^{11}	10^6
Strontium					
Sr-85	1	10^6	10^2	10^{11}	10^7
Sr-85m	10^2	10^7	10^2	10^{13}	10^8
Sr-87m	10^2	10^6	10^2	10^{13}	10^7
Sr-89	10^3	10^6	10^3	10^{10}	10^7
Sr-90+	1	10^4	10^2	10^9	10^5
Sr-91+	10	10^5	10	10^{12}	10^6
Sr-92	10	10^6	10	10^{12}	10^7
Yttrium					
Y-90	10^3	10^5	10^3	10^{11}	10^6
Y-91	10^2	10^6	10^3	10^{10}	10^7
Y-91m	10^2	10^6	10^2	10^{13}	10^7
Y-92	10^2	10^5	10^2	10^{12}	10^6
Y-93	10^2	10^5	10^2	10^{12}	10^6
Zirconium					
Zr-93+	10	10^7	10^3	10^9	10^8

Schedule 7

1	2	3	4	5	6
Radionuclide name, symbol, isotope	Concentration for: Notification (any amount of radioactive material); Registration (amounts of radioactive material that exceed 1,000kg) Regulation 5(1) and Schedule 1, paragraph 1(a); regulation 6(2)(f) (Bq/g)	Quantity for Notification Regulation 5(1) and Schedule 1, paragraph 1(b) (Bq)	Concentration for Registration (amounts of radioactive material that do not exceed 1,000kg) Regulation 6(2)(e) (Bq/g)	Quantity for notification of occurrences Regulation 31(1) (Bq)	Quantity for notification of occurrences Regulation 31(3) (Bq)
Zr-95+	1	10^6	10	10^{10}	10^7
Zr-97+	10	10^5	10	10^{11}	10^6
Niobium					
Nb-93m	10	10^7	10^4	10^{11}	10^8
Nb-94	0.1	10^6	10	10^9	10^7
Nb-95	1	10^6	10	10^{11}	10^7
Nb-97+	10	10^6	10	10^{13}	10^7
Nb-98	10	10^5	10	10^{13}	10^6
Molybdenum					
Mo-90	10	10^6	10	10^{12}	10^7
Mo-93	10	10^8	10^3	10^{11}	10^9
Mo-99+	10	10^6	10^2	10^{11}	10^7
Mo-101+	10	10^6	10	10^{13}	10^7
Technetium					
Tc-96	1	10^6	10	10^{11}	10^7
Tc-96m	10^3	10^7	10^3	10^{14}	10^8
Tc-97	10	10^8	10^3	10^{12}	10^9
Tc-97m	10^2	10^7	10^3	10^{10}	10^8
Tc-99	1	10^7	10^4	10^{10}	10^8
Tc-99m	10^2	10^7	10^2	10^{13}	10^8
Ruthenium					
Ru-97	10	10^7	10^2	10^{12}	10^8
Ru-103+	1	10^6	10^2	10^{10}	10^7
Ru-105+	10	10^6	10	10^{12}	10^7
Ru-106+	0.1	10^5	10^2	10^9	10^6
Rhodium					
Rh-103m	10^4	10^8	10^4	10^{15}	10^9
Rh-105	10^2	10^7	10^2	10^{12}	10^8
Palladium					
Pd-103+	10^3	10^8	10^3	10^{11}	10^9
Pd-109+	10^2	10^6	10^3	10^{12}	10^7
Silver					
Ag-105	1	10^6	10^2	10^{11}	10^7
Ag-108m+	0.1	10^6	10	10^{10}	10^7

Schedule 7

1	2	3	4	5	6
Radionuclide name, symbol, isotope	Concentration for: Notification (any amount of radioactive material); Registration (amounts of radioactive material that exceed 1,000kg) Regulation 5(1) and Schedule 1, paragraph 1(a); regulation 6(2)(f) (Bq/g)	Quantity for Notification Regulation 5(1) and Schedule 1, paragraph 1(b) (Bq)	Concentration for Registration (amounts of radioactive material that do not exceed 1,000kg) Regulation 6(2)(e) (Bq/g)	Quantity for notification of occurrences Regulation 31(1) (Bq)	Quantity for notification of occurrences Regulation 31(3) (Bq)
Ag-110m+	0.1	10^6	10	10^{10}	10^7
Ag-111	10^2	10^6	10^3	10^{11}	10^7
Cadmium					
Cd-109+	1	10^6	10^4	10^{10}	10^7
Cd-115+	10	10^6	10^2	10^{11}	10^7
Cd-115m+	10^2	10^6	10^3	10^{10}	10^7
Indium					
In-111	10	10^6	10^2	10^{11}	10^7
In-113m	10^2	10^6	10^2	10^{13}	10^7
In-114m+	10	10^6	10^2	10^{10}	10^7
In-115m	10^2	10^6	10^2	10^{13}	10^7
Tin					
Sn-113+	1	10^7	10^3	10^{11}	10^8
Sn-125	10	10^5	10^2	10^{10}	10^6
Antimony					
Sb-122	10	10^4	10^2	10^{11}	10^5
Sb-124	1	10^6	10	10^{10}	10^7
Sb-125+	0.1	10^6	10^2	10^{10}	10^7
Tellurium					
Te-123m	1	10^7	10^2	10^{10}	10^8
Te-125m	10^3	10^7	10^3	10^{10}	10^8
Te-127	10^3	10^6	10^3	10^{12}	10^7
Te-127m+	10	10^7	10^3	10^{10}	10^8
Te-129	10^2	10^6	10^2	10^{14}	10^7
Te-129m+	10	10^6	10^3	10^{10}	10^7
Te-131	10^2	10^5	10^2	10^{14}	10^6
Te-131m+	10	10^6	10	10^{11}	10^7
Te-132+	1	10^7	10^2	10^{11}	10^8
Te-133	10	10^5	10	10^{14}	10^6
Te-133m	10	10^5	10	10^{13}	10^6
Te-134	10	10^6	10	10^{13}	10^7
Iodine					
I-123	10^2	10^7	10^2	10^{12}	10^8
I-125	10^2	10^6	10^3	10^{10}	10^7

Schedule 7

1	2	3	4	5	6
Radionuclide name, symbol, isotope	Concentration for: Notification (any amount of radioactive material); Registration (amounts of radioactive material that exceed 1,000kg) Regulation 5(1) and Schedule 1, paragraph 1(a); regulation 6(2)(f) (Bq/g)	Quantity for Notification Regulation 5(1) and Schedule 1, paragraph 1(b) (Bq)	Concentration for Registration (amounts of radioactive material that do not exceed 1,000kg) Regulation 6(2)(e) (Bq/g)	Quantity for notification of occurrences Regulation 31(1) (Bq)	Quantity for notification of occurrences Regulation 31(3) (Bq)
I-126	10	10^6	10^2	10^{10}	10^7
I-129	0.01	10^5	10^2	10^9	10^6
I-130	10	10^6	10	10^{11}	10^7
I-131	10	10^6	10^2	10^{10}	10^7
I-132	10	10^5	10	10^{12}	10^6
I-133	10	10^6	10	10^{11}	10^7
I-134	10	10^5	10	10^{13}	10^6
I-135	10	10^6	10	10^{12}	10^7
Xenon					
Xe-131m	0.01	10^4	10^4	10^{11}	
Xe-133	0.01	10^4	10^3	10^{11}	
Xe-135	0.01	10^{10}	10^3	10^{10}	
Caesium					
Cs-129	10	10^5	10^2	10^{12}	10^6
Cs-131	10^3	10^6	10^3	10^{12}	10^7
Cs-132	10	10^5	10	10^{11}	10^6
Cs-134	0.1	10^4	10	10^{10}	10^5
Cs-134m	10^3	10^5	10^3	10^{14}	10^6
Cs-135	10^2	10^7	10^4	10^{11}	10^8
Cs-136	1	10^5	10	10^{10}	10^6
Cs-137+	0.1	10^4	10	10^{10}	10^5
Cs-138	10	10^4	10	10^{13}	10^5
Barium					
Ba-131	10	10^6	10^2	10^{11}	10^7
Ba-140+	1	10^5	10	10^{11}	10^6
Lanthanum					
La-140	1	10^5	10	10^{11}	10^6
Cerium					
Ce-139	1	10^6	10^2	10^{11}	10^7
Ce-141	10^2	10^7	10^2	10^{10}	10^8
Ce-143	10	10^6	10^2	10^{11}	10^7
Ce-144+	10	10^5	10^2	10^9	10^6

Schedule 7

1	2	3	4	5	6
Radionuclide name, symbol, isotope	Concentration for: Notification (any amount of radioactive material); Registration (amounts of radioactive material that exceed 1,000kg)	Quantity for Notification	Concentration for Registration (amounts of radioactive material that do not exceed 1,000kg)	Quantity for notification of occurrences	Quantity for notification of occurrences
	Regulation 5(1) and Schedule 1, paragraph 1(a); regulation 6(2)(f)	Regulation 5(1) and Schedule 1, paragraph 1(b)	Regulation 6(2)(e)	Regulation 31(1)	Regulation 31(3)
	(Bq/g)	(Bq)	(Bq/g)	(Bq)	(Bq)
Praseodymium					
Pr-142	10^2	10^5	10^2	10^{12}	10^6
Pr-143	10^3	10^6	10^4	10^{11}	10^7
Neodymium					
Nd-147	10^2	10^6	10^2	10^{11}	10^7
Nd-149	10^2	10^6	10^2	10^{13}	10^7
Promethium					
Pm-147	10^3	10^7	10^4	10^{10}	10^8
Pm-149	10^3	10^6	10^3	10^{11}	10^7
Samarium					
Sm-151	10^3	10^8	10^4	10^{10}	10^9
Sm-153	10^2	10^6	10^2	10^{11}	10^7
Europium					
Eu-152	0.1	10^6	10	10^9	10^7
Eu-152m	10^2	10^6	10^2	10^{12}	10^7
Eu-154	0.1	10^6	10	10^9	10^7
Eu-155	1	10^7	10^2	10^{10}	10^8
Gadolinium					
Gd-153	10	10^7	10^2	10^{10}	10^8
Gd-159	10^2	10^6	10^3	10^{12}	10^7
Terbium					
Tb-160	1	10^6	1	10^{10}	10^7
Dysprosium					
Dy-165	10^3	10^6	10^3	10^{13}	10^7
Dy-166	10^2	10^6	10^3	10^{11}	10^7
Holmium					
Ho-166	10^2	10^5	10^3	10^{11}	10^6
Erbium					
Er-169	10^3	10^7	10^4	10^{11}	10^8
Er-171	10^2	10^6	10^2	10^{12}	10^7
Thulium					
Tm-170	10^2	10^6	10^3	10^{10}	10^7
Tm-171	10^3	10^8	10^4	10^{11}	10^9

Schedule 7

1	2	3	4	5	6
Radionuclide name, symbol, isotope	Concentration for: Notification (any amount of radioactive material); Registration (amounts of radioactive material that exceed 1,000kg) Regulation 5(1) and Schedule 1, paragraph 1(a); regulation 6(2)(f) (Bq/g)	Quantity for Notification Regulation 5(1) and Schedule 1, paragraph 1(b) (Bq)	Concentration for Registration (amounts of radioactive material that do not exceed 1,000kg) Regulation 6(2)(e) (Bq/g)	Quantity for notification of occurrences Regulation 31(1) (Bq)	Quantity for notification of occurrences Regulation 31(3) (Bq)
Ytterbium					
Yb-175	10^2	10^7	10^3	10^{11}	10^8
Lutetium					
Lu-177	10^2	10^7	10^3	10^{11}	10^8
Hafnium					
Hf-181	1	10^6	10	10^{10}	10^7
Tantalum					
Ta-182	0.1	10^4	10	10^{10}	10^5
Tungsten					
W-181	10	10^7	10^3	10^{12}	10^8
W-185	10^3	10^7	10^4	10^{11}	10^8
W-187	10	10^6	10^2	10^{12}	10^7
Rhenium					
Re-186	10^3	10^6	10^3	10^{11}	10^7
Re-188	10^2	10^5	10^2	10^{12}	10^6
Osmium					
Os-185	1	10^6	10	10^{11}	10^7
Os-191	10^2	10^7	10^2	10^{11}	10^8
Os-191m	10^3	10^7	10^3	10^{12}	10^8
Os-193	10^2	10^6	10^2	10^{11}	10^7
Iridium					
Ir-190	1	10^6	10	10^{10}	10^7
Ir-192	1	10^4	10	10^{10}	10^5
Ir-194	10^2	10^5	10^2	10^{11}	10^6
Platinum					
Pt-191	10	10^6	10^2	10^{11}	10^7
Pt-193m	10^3	10^7	10^3	10^{12}	10^8
Pt-197	10	10^6	10^3	10^{12}	10^7
Pt-197m	10^2	10^6	10^2	10^{14}	10^7
Gold					
Au-198	10	10^6	10^2	10^{11}	10^7
Au-199	10^2	10^6	10^2	10^{11}	10^7

Schedule 7

1	2	3	4	5	6
Radionuclide name, symbol, isotope	Concentration for: Notification (any amount of radioactive material); Registration (amounts of radioactive material that exceed 1,000kg) Regulation 5(1) and Schedule 1, paragraph 1(a); regulation 6(2)(f) (Bq/g)	Quantity for Notification Regulation 5(1) and Schedule 1, paragraph 1(b) (Bq)	Concentration for Registration (amounts of radioactive material that do not exceed 1,000kg) Regulation 6(2)(e) (Bq/g)	Quantity for notification of occurrences Regulation 31(1) (Bq)	Quantity for notification of occurrences Regulation 31(3) (Bq)
Mercury					
Hg-197	10^2	10^7	10^2	10^{12}	10^8
Hg-197m	10^2	10^6	10^2	10^{12}	10^7
Hg-203	10	10^5	10^2	10^{11}	10^6
Thallium					
Tl-200	10	10^6	10	10^{11}	10^7
Tl-201	10^2	10^6	10^2	10^{12}	10^7
Tl-202	10	10^6	10^2	10^{11}	10^7
Tl-204	1	10^4	10^4	10^{11}	10^5
Lead					
Pb-203	10	10^6	10^2	10^{12}	10^7
Pb-210+	0.01	10^4	10	10^8	10^5
Pb-212+	1	10^5	10	10^{10}	10^6
Bismuth					
Bi-206	1	10^5	10	10^{10}	10^6
Bi-207	0.1	10^6	10	10^{10}	10^7
Bi-210	10	10^6	10^3	10^9	10^7
Bi-212+	1	10^5	10	10^{11}	10^6
Polonium					
Po-203	10	10^6	10	10^{13}	10^7
Po-205	10	10^6	10	10^{12}	10^7
Po-207	10	10^6	10	10^{12}	10^7
Po-210	0.01	10^4	10	10^7	10^5
Astatine					
At-211	10^3	10^7	10^3	10^{10}	10^8
Radon					
Rn-220+	0.01	10^7	10^4	10^8	10^8
Rn-222+	0.01	10^8	10	10^9	10^9
Radium					
Ra-223+	1	10^5	10^2	10^7	10^6
Ra-224+	1	10^5	10	10^8	10^6
Ra-225	10	10^5	10^2	10^7	10^6
Ra-226+	0.01	10^4	10	10^7	10^5

Schedule 7

1	2	3	4	5	6
Radionuclide name, symbol, isotope	Concentration for: Notification (any amount of radioactive material); Registration (amounts of radioactive material that exceed 1,000kg) Regulation 5(1) and Schedule 1, paragraph 1(a); regulation 6(2)(f) (Bq/g)	Quantity for Notification Regulation 5(1) and Schedule 1, paragraph 1(b) (Bq)	Concentration for Registration (amounts of radioactive material that do not exceed 1,000kg) Regulation 6(2)(e) (Bq/g)	Quantity for notification of occurrences Regulation 31(1) (Bq)	Quantity for notification of occurrences Regulation 31(3) (Bq)
Ra-227	10^2	10^6	10^2	10^{13}	10^7
Ra-228+	0.01	10^5	10	10^8	10^6
Actinium					
Ac-228	1	10^6	10	10^{10}	10^7
Thorium					
Th-226+	10^3	10^7	10^3	10^{11}	10^8
Th-227	1	10^4	10	10^7	10^5
Th-228+	0.1	10^4	1	10^6	10^5
Th-229+	0.1	10^3	1	10^6	10^4
Th-230	0.1	10^4	1	10^6	10^5
Th-231	10^2	10^7	10^3	10^{12}	10^8
Th-232	0.01	10^4	10	10^6	10^5
Th-234+	10	10^5	10^3	10^{10}	10^6
Protactinium					
Pa-230	10	10^6	10	10^8	10^7
Pa-231	0.01	10^3	1	10^6	10^4
Pa-233	10	10^7	10^2	10^{10}	10^8
Uranium					
U-230+	10	10^5	10	10^7	10^6
U-231	10^2	10^7	10^2	10^{11}	10^8
U-232+	0.1	10^3	1	10^6	10^4
U-233	1	10^4	10	10^7	10^5
U-234	1	10^4	10	10^7	10^5
U-235+	1	10^4	10	10^7	10^5
U-236	10	10^4	10	10^7	10^5
U-237	10^2	10^6	10^2	10^{11}	10^7
U-238+	1	10^4	10	10^7	10^5
U-239	10^2	10^6	10^2	10^{14}	10^7
U-240	0.01	10^7	10^3	10^{12}	10^8
U-240+	10^2	10^6	10	10^{11}	10^7
Neptunium					
Np-237+	1	10^3	1	10^7	10^4
Np-239	10^2	10^7	10^2	10^{11}	10^8
Np-240	10	10^6	10	10^{13}	10^7

Work with ionising radiation

Schedule 7

1	2	3	4	5	6
Radionuclide name, symbol, isotope	Concentration for: Notification (any amount of radioactive material); Registration (amounts of radioactive material that exceed 1,000kg) Regulation 5(1) and Schedule 1, paragraph 1(a); regulation 6(2)(f) (Bq/g)	Quantity for Notification Regulation 5(1) and Schedule 1, paragraph 1(b) (Bq)	Concentration for Registration (amounts of radioactive material that do not exceed 1,000kg) Regulation 6(2)(e) (Bq/g)	Quantity for notification of occurrences Regulation 31(1) (Bq)	Quantity for notification of occurrences Regulation 31(3) (Bq)
Plutonium					
Pu-234	10^2	10^7	10^2	10^{10}	10^8
Pu-235	10^2	10^7	10^2	10^{14}	10^8
Pu-236	1	10^4	10	10^7	10^5
Pu-237	10^2	10^7	10^3	10^{11}	10^8
Pu-238	0.1	10^4	1	10^6	10^5
Pu-239	0.1	10^4	1	10^6	10^5
Pu-240	0.1	10^3	1	10^6	10^4
Pu-241	10	10^5	10^2	10^8	10^6
Pu-242	0.1	10^4	1	10^6	10^5
Pu-243	10^3	10^7	10^3	10^{13}	10^8
Pu-244+	0.1	10^4	1	10^6	10^5
Americium					
Am-241	0.1	10^4	1	10^6	10^5
Am-242	10^3	10^6	10^3	10^{10}	10^7
Am-242m+	0.1	10^4	1	10^6	10^5
Am-243+	0.1	10^3	1	10^6	10^4
Curium					
Cm-242	10	10^5	10^2	10^7	10^6
Cm-243	1	10^4	1	10^7	10^5
Cm-244	1	10^4	10	10^7	10^5
Cm-245	0.1	10^3	1	10^6	10^4
Cm-246	0.1	10^3	1	10^6	10^4
Cm-247+	0.1	10^4	1	10^6	10^5
Cm-248	0.1	10^3	1	10^6	10^4
Berkelium					
Bk-249	10^2	10^6	10^3	10^9	10^7
Californium					
Cf-246	10^3	10^6	10^3	10^9	10^7
Cf-248	1	10^4	10	10^7	10^5
Cf-249	0.1	10^3	1	10^6	10^4
Cf-250	1	10^4	10	10^6	10^5
Cf-251	0.1	10^3	1	10^6	10^4
Cf-252	1	10^4	10	10^7	10^5

Schedule 7

1	2	3	4	5	6
Radionuclide name, symbol, isotope	Concentration for: Notification (any amount of radioactive material); Registration (amounts of radioactive material that exceed 1,000kg)	Quantity for Notification	Concentration for Registration (amounts of radioactive material that do not exceed 1,000kg)	Quantity for notification of occurrences	Quantity for notification of occurrences
	Regulation 5(1) and Schedule 1, paragraph 1(a); regulation 6(2)(f)	Regulation 5(1) and Schedule 1, paragraph 1(b)	Regulation 6(2)(e)	Regulation 31(1)	Regulation 31(3)
	(Bq/g)	(Bq)	(Bq/g)	(Bq)	(Bq)
Cf-253	10^2	10^5	10^2	10^8	10^6
Cf-254	1	10^3	1	10^7	10^4
Einsteinium					
Es-253	10^2	10^5	10^2	10^8	10^6
Es-254+	0.1	10^4	10	10^7	10^5
Es-254m+	10	10^6	10^2	10^9	10^7
Fermium					
Fm-254	10^4	10^7	10^4	10^{10}	10^8
Fm-255	10^2	10^6	10^3	10^9	10^7
Other radionuclides not listed above (see Note 1)					
	0.01	10^3	0.1	10^5	10^4

Note 1

In the case of radionuclides not specified elsewhere in this Part, the quantities specified in this entry are to be used unless the Executive has approved some other quantity for that radionuclide.

Note 2

Nuclides carrying the suffix "+" in the above table represent parent nuclides and their progeny as listed in the table below. The dose contributions for those progeny are taken into account in the dose calculation (thus requiring only the exemption level of the parent radionuclide to be considered).

[1] Potassium salts in quantities less than 1,000kg are exempt.

List of parent nuclides and their progeny as referred to in Note 2 above

Parent radionuclide	Progeny
Fe-52	Mn-52m
Zn-69m	Zn-69
Ge-68	Ga-68
Sr-90	Y-90
Sr-91	Y-91m
Zr-93	Nb-93m
Zr-95	Nb-95
Zr-97	Nb-97m, Nb-97
Nb-97	Nb-97m
Mo-99	Tc-99m
Mo-101	Tc-101

Schedule 7

Parent radionuclide	Progeny
Ru-103	Rh-103m
Ru-105	Rh-105m
Ru-106	Rh-106
Pd-103	Rh-103m
Pd-109	Ag-109m
Ag-108m	Ag-108
Ag-110m	Ag-110
Cd-109	Ag-109m
Cd-115	In-115m
Cd-115m	In-115m
In-114m	In-114
Sn-113	In-113m
Sb-125	Te-125m
Te-127m	Te-127
Te-129m	Te-129
Te-131m	Te-131
Te-132	I-132
Cs-137	Ba-137m
Ba-140	La-140
Ce-144	Pr-144, Pr-144m
Pb-210	Bi-210, Po-210
Pb-212	Bi-212, Tl-208, Po-212
Bi-212	Tl-208, Po-212
Rn-220	Po-216
Rn-222	Po-218, Pb-214, Bi-214, Po-214
Ra-223	Rn-219, Po-215, Pb-211, Bi-211, Tl-207
Ra-224	Rn-220, Po-216, Pb-212, Bi-212, Tl-208, Po-212
Ra-226	Rn-222, Po-218, Pb-214, Bi-214, Po-214, Pb-210, Bi-210, Po-210
Ra-228	Ac-228
Th-226	Ra-222, Rn-218, Po-214
Th-228	Ra-224, Rn-220, Po-216, Pb-212, Bi-212, Tl-208, Po-212
Th-229	Ra-225, Ac-225, Fr-221, At-217, Bi-213, Po-213, Pb-209
Th-234	Pa-234m
U-230	Th-226, Ra-222, Rn-218, Po-214
U-232	Th-228, Ra-224, Rn-220, Po-216, Pb-212, Bi-212, Tl-208, Po-212
U-235	Th-231
U-238	Th-234, Pa-234m
U-240	Np-240m, Np-240
Np-237	Pa-233
Pu-244	U-240, Np-240m, Np-240
Am-242m	Am-242, Np-238
Am-243	Np-239
Cm-247	Pu-243
Es-254	Bk-250
Es-254m	Fm-254

Schedule	7

Part 2 Regulations 2(4), 5(1) 6(2) and Schedule 1

Table of naturally occurring radionuclides (which are not processed for their radioactive, fissile or fertile properties)

Values for exemption from notification and registration for naturally occurring radionuclides in solid materials (which are not processed for their radioactive, fissile or fertile properties), which apply whether or not the radionuclide is in secular equilibrium with its progeny.

1	2	3	4
Radionuclide name, symbol, isotope	Concentration for: Notification (any amount of radioactive material); Registration (amounts of radioactive material that exceed 1,000kg) Regulation 5(1) and Schedule 1, paragraph 1(a); regulation 6(2)(f) (Bq/g)	Quantity for Notification Regulation 5(1) and Schedule 1, paragraph 1(b) (Bq)	Concentration for Registration (amounts of radioactive material that do not exceed 1,000kg) Regulation 6(2)(e) (Bq/g)
K-40[(1)]	10	10^6	10^2
Rb-87	1	10^7	10^4
Pb-210+	1	10^4	10
Po-210	1	10^4	10
Ra-226+	1	10^4	10
Ra-228+	1	10^5	10
Th-228+	1	10^4	1
Th-232 sec	1	10^3	1
U-238 sec	1	10^3	1

Note

Nuclides carrying the suffix "+" in the above table represent parent nuclides and their progeny as listed in the table below. The dose contributions of those progeny are taken into account in the dose calculation (thus requiring only the exemption level of the parent radionuclide to be considered).

[(1)] *Potassium salts in quantities less than 1,000kg are exempt.*

List of parent nuclides and their progeny as referred to in the Note above

Parent radionuclide	Progeny
Pb-210	Bi-210, Po-210
Ra-226	Rn-222, Po-218, Pb-214, Bi-214, Po-214, Pb-210, Bi-210, Po-210
Ra-228	Ac-228
Th-228	Ra-224, Rn-220, Po-216, Pb-212, Bi-212, Tl-208, Po-212

Schedule	7

Part 3 Regulation 2(4)

Quantity and concentration ratios for more than one radionuclide

(1)   For the purpose of Regulation 2(4) –

(a)   the quantity ratio for more than one radionuclide is the sum of the quotients of the quantity of a radionuclide present Q_p divided by the quantity of that radionuclide specified in the appropriate entry in Parts 1, 2 or 4 of this Schedule Q_{lim}, namely—

$$\sum \frac{Q_p}{Q_{lim}}$$

(b)   the concentration ratio for more than one radionuclide is the sum of the quotients of the concentration of a radionuclide present C_p divided by the concentration of that radionuclide specified in the appropriate entry in Parts 1 or 2 of this Schedule C_{lim}, namely –

$$\sum \frac{C_p}{C_{lim}}$$

(2)   In any case where the isotopic composition of a radioactive substance is not known or is only partially known, the quantity or concentration ratio for that substance is to be calculated by using the values specified in the appropriate column in Part 1 of this Schedule for "other radionuclides not listed above" for any radionuclide that has not been identified or where the quantity or concentration of a radionuclide is uncertain, unless the employer can show that the use of some other value is appropriate in the circumstances of a particular case, when the employer may use that value.

Schedule	7

Part 4 Regulations 2(1) and 2(4)

Table of quantities of radioactive material defining high-activity sealed sources

For radionuclides not listed in the table below, the relevant quantity value is the same as the D-value defined in section 2 Table 1 of the IAEA publication: Dangerous quantities of radioactive material (D-values), (EPR-D-VALUES 2006)

Radionuclide	Quantity (Bq)
Co-60	3×10^{10}
Se-75	2×10^{11}
Sr-90 (Y-90)	1×10^{12}
Cs-137	1×10^{11}
Pm-147	4×10^{13}
Gd-153	1×10^{12}
Tm-170	2×10^{13}
Yb-169	3×10^{11}
Ir-192	8×10^{10}
Ra-226	4×10^{10}
Pu-238	6×10^{10}
Pu-239/Be-9[*]	6×10^{10}
Am-241	6×10^{10}
Am-241/Be-9[*]	6×10^{10}
Cm-244	5×10^{10}
Cf-252	2×10^{10}

[*] The activity given is that of the alpha-emitting radionuclide.

Schedule 8 Transitional provisions and savings

Schedule	8

Regulation 41

(1)   (1) In this Schedule –

"the 1999 Regulations" means the Ionising Radiations Regulations 1999;

"restated provision" means any provision of these Regulations so far as it corresponds (with or without modification) to a provision of the 1999 Regulations;

"superseded provision" means any provision of the 1999 Regulations as it has effect immediately before 1st January 2018 so far as it corresponds (with or without modification) to a provision of these Regulations.

(2)   In this Schedule references to things done include references to things omitted to be done.

(2)   (1) Any thing done, or having effect as if done, under or for the purposes of any superseded provision, if effective immediately before 1st January 2018, has effect, so far as is required for continuing its effect on and after that date, as if done under or for the purposes of the corresponding restated provision.

(2)   Paragraph (1) does not apply in relation to an authorisation granted or notification made under the 1999 Regulations.

(3)   The specific provisions in paragraphs 3 to 10 are not to be taken to affect the generality of paragraph (1).

(3)   Where on or before 5th February 2018 an employer commences work in respect of which a notification is required under regulation 5(2), it will be sufficient compliance with that regulation if the employer notifies the appropriate authority and provides the particulars required under regulation 5(2) on or before 5th February 2018.

(4)   In paragraph 3 "appropriate authority" has the same meaning as set out in regulation 5(6).

(5)   Where on or before 5th February 2018 a person carries out a registrable practice (within the meaning of regulation 6(1)) it will be sufficient compliance with regulation 6(3) if the person completes the registration procedure under that regulation on or before 5th February 2018.

(6)   A person who carries out a practice requiring consent under regulation 7 on or before 5th February 2018 is deemed to have been granted consent to carry out that practice under regulation 7(2) until 5th February 2018.

(7)   Where an employer has, in respect of an employee, applied the dose limits set out in paragraphs 9 to 11 of Schedule 4 to the 1999 Regulations in accordance with the requirements of regulation 11(2) of those Regulations and those dose limits have effect immediately before 1st January 2018, the appropriate authority is deemed to

Schedule 8

have approved, for the purposes of regulation 12(2) of these Regulations, the application of the dose limits, in respect of that employee, set out in paragraphs 9 to 11 of Schedule 3 to these Regulations.

(8)   In paragraph 7 –

(a)   "appropriate authority" has the same meaning as set out in regulation 12(4);

(b)   the deemed approval granted by that paragraph is valid until the end of 5th February 2018.

(9)   A radiation passbook approved for the purposes of the 1999 Regulations and issued on or before 30th April 2018 in respect of a classified outside worker employed by an employer in Great Britain and which was at that date valid remains valid for such time as the worker to whom the passbook relates continues to be employed by the same employer.

(10)   Where a superseded provision provides a period of time within which an aggrieved person may apply for a decision to be reviewed, that period of time continues to apply on and after 1st January 2018 in relation to any decision notified to the aggrieved person before 1 January 2018.

Schedule 9 Modifications

Schedule	9

Regulation 42

The Employment Act 1989
1. In Schedule 1 to the Employment Act 1989, omit "Paragraphs 5 and 11 of Schedule 4 to the Ionising Radiations Regulations 1999 [SI 1999/3232]".

The Employment Rights Act 1996
2. In section 64(3) of the Employment Rights Act 1996, for "Regulation 24 of the Ionising Radiations Regulations 1999 [SI 1999/3232]" substitute "Regulation 25 of the Ionising Radiations Regulation 2017 [SI 2017/1075]".

The Personal Protective Equipment at Work Regulations 1992
3. In regulation 3(3)(b) of the Personal Protective Equipment at Work Regulations 1992, for "the Ionising Radiations Regulations 1999 [SI 1999/3232]" substitute "the Ionising Radiations Regulations 2017 [SI 2017/1075]".

The Health and Safety (Enforcing Authority) Regulations 1998
4. (1) The Health and Safety (Enforcing Authority) Regulations 1998are amended as follows.

(2) In regulation 2(1), in the definition of "ionising radiation", for "the Ionising Radiations Regulations 1999 [SI 1999/3232]" substitute "the Ionising Radiations Regulations 2017 [SI 2017/1075]".

(3) In regulation 4A (the Office for Nuclear Regulation)—

    (a) in paragraph (2), for sub-paragraph (a) substitute—
      "(a) the provisions of the Ionising Radiations Regulations 2017 in so far as they apply—
        (i) in relation to the civil transport of radioactive material by road, railway or inland waterway; and
        (ii) to premises which are or are on a nuclear warship site;
      (aa) the provisions of the Radiation (Emergency Preparedness and Public Information) Regulations 2001 in so far as they apply to premises which are or are on a nuclear warship site;"
    (b) for paragraph (3) substitute—
      "(3) For the purposes of—
      (a) paragraph (2)(a)—
        (i) "civil transport" means transport otherwise than for the purposes of the department of the Secretary of State with responsibility for defence;
        (ii) "radioactive material" has the same meaning as given in regulation 2(1) of the Ionising Radiations Regulations 2017 [S.I. 2017/1075];

Schedule 9

(iii)   the transport of material begins with any preparatory process (such as packaging) and continues until the material has been unloaded at its destination;

(b)   paragraphs (2)(a) and (aa) "premises" includes a nuclear powered warship during any period it is berthed or anchored at a nuclear warship site."

(4)   In Schedule 2—

(a)   in paragraph 4(d), for "Schedule 1 of the Ionising Radiations Regulations 1999 [SI 1999/3232]" substitute "Schedule 1 to the Ionising Radiations Regulations 2017 [SI 2017/1075]";

(b)   in paragraph 5, for "the Ionising Radiations Regulations 1999 [SI 1999/3232]" substitute "the Ionising Radiations Regulations 2017 [SI 2017/1075]".

The Radiation (Emergency Preparedness and Public Information) Regulations 2001

5.   (1)   The Radiation (Emergency Preparedness and Public Information) Regulations 2001 are amended as follows.

(2)   In regulation 2(1)—

(a)   for the definition of "the 1999 Regulations" substitute— "the 2017 Regulations" means the "Ionising Radiations Regulations 2017";

(b)   in the definition of "approved dosimetry service", for "the 1999 Regulations" substitute "the 2017 Regulations";

(c)   in the definition of "dose assessment", for "regulation 21 of the 1999 Regulations" substitute "regulation 22 of the 2017 Regulations";

(d)   in the definition of "dose record", for "regulation 21 of the 1999 Regulations" substitute "regulation 22 of the 2017 Regulations";

(e)   in the definition of "emergency exposure", for "Schedule 4 to the 1999 Regulations" substitute "Schedule 3 to the 2017 Regulations";

(f)   in the definition of "medical surveillance", for "regulation 24 of the 1999 Regulations" substitute "regulation 25 of the 2017 Regulations".

(3)   In regulation 4(3), for "regulation 7 (Prior risk assessment etc) of the 1999 Regulations" substitute "regulation 8 (Radiation risk assessments) of the 2017 Regulations".

(4)   In regulations 7(7)(b) and 8(8)(b), for "regulation 21 of the 1999 Regulations" substitute "regulation 22 of the 2017 Regulations" in each case.

(5)   In regulation 15, for "regulation 11 of the 1999 Regulations" substitute "regulation 12 of the 2017 Regulations".

(6)   In Schedule 11 omit paragraphs 2 to 9.

The High-activity Sealed Radioactive Sources and Orphan Sources Regulations 2005

6.   In the High-activity Sealed Radioactive Sources and Orphan Sources Regulations 2005, omit regulation 19.

Schedule	**9**

The Health and Safety (Enforcing Authority for Railways and Other Guided Transport Systems) Regulations 2006

7.   (1)   The Health and Safety (Enforcing Authority for Railways and Other Guided Transport Systems) Regulations 2006 are amended as follows.

(2)   In regulation 3 (enforcing authority)—

(a)   after paragraph (4) insert—
"(4A) The Office of Rail and Road has no responsibility for the enforcement of the Ionising Radiations Regulations 2017."

(b)   in paragraph (5)—
(i)   for "regulation 93(4)" substitute "regulation 32(4)";
(ii)   for "2007 (defence and enforcement)" substitute "2009 (enforcement)".

The Legislative Reform (Health and Safety Executive) Order 2008

8.   Schedule 3 to the Legislative Reform (Health and Safety Executive) Order 2008 omit the entry relating to the Ionising Radiations Regulations 1999.

The REACH Enforcement Regulations 2008

9.   In Part 3 of Schedule 3 to the REACH Enforcement Regulations 2008—

(a)   in paragraph 1(g)(i), for "the Ionising Radiations Regulations 1999" substitute "the Ionising Radiations Regulations 2017";
(b)   in paragraph 3, for "the Ionising Radiations Regulations" substitute "the Ionising Radiations Regulations 2017".

The Carriage of Dangerous Goods and Use of Transportable Pressure Equipment Regulations 2009

10.   (1)   Schedule 2 to the Carriage of Dangerous Goods and Use of Transportable Pressure Equipment Regulations 2009 is amended as follows.

(2)   In paragraph 3(1)—

(a)   for "regulation 20 of the Ionising Radiations Regulations 1999 ("the 1999 Regulations")" substitute "regulation 21 of the Ionising Radiations Regulations 2017 ("the 2017 Regulations")";
(b)   for "regulations 21 to 26 of the 1999 Regulations" substitute "regulations 22 to 27 of the 2017 Regulations".

(3)   In paragraph 3(2), for "paragraph 1, 2, 6, 7 or 8 of Schedule 4 (Dose Limits) to the 1999 Regulations" substitute "paragraphs 1, 2, 5, 6 or 7 of Schedule 3 (Dose limits) to the 2017 Regulations".

(4)   In paragraph 3(3), for "Schedule 4 to the 1999 Regulations" substitute "Schedule 3 to the 2017 Regulations".

(5)   In paragraph 4(2)(c), for "Schedule 4 to the Ionising Radiations Regulations 1999" substitute "Schedule 3 to the Ionising Radiations Regulations 2017".

The Environmental Permitting (England and Wales) Regulations 2010

11.   In Part 2 of Schedule 26 to the Environmental Permitting (England and Wales) Regulations 2010, omit paragraph 15 (Ionising Radiations Regulations 1999).

Schedule	9

The Natural Resources Body for Wales (Functions) Order 2013
12.   In Schedule 4 to the Natural Resources Body for Wales (Functions) Order 2013, omit paragraph 113 (Ionising Radiations Regulations 1999).

The Reporting of Injuries, Diseases and Dangerous Occurrences Regulations 2013
13.   (1)   The Reporting of Injuries, Diseases and Dangerous Occurrences Regulations 2013 are amended as follows.

(2)   In regulation 14(6)(e), for "the Ionising Radiations Regulations 1999" substitute "the Ionising Radiations Regulations 2017".

(3)   In Schedule 4, Table 1, omit the entry relating to the Ionising Radiations Regulations 1999.

The Construction (Design and Management) Regulations 2015
14.   In paragraph 3 of Schedule 3 to the Construction (Design and Management) Regulations 2015, for "regulation 16 of the Ionising Radiations Regulations 1999" substitute "regulation 17 of the Ionising Radiations Regulations 2017".

The Infrastructure Planning (Interested Parties and Miscellaneous Prescribed Provisions) Regulations 2015
15.   (1)   The Infrastructure Planning (Interested Parties and Miscellaneous Prescribed Provisions) Regulations 2015 are amended as follows.

(2)   In Part 2 of Schedule 2—

(a)   in column 1, for "Ionising Radiations Regulations 1999" substitute "Ionising Radiations Regulations 2017";
(b)   in column 2, for the corresponding entry, for "Authorisation under regulation 5 (authorisation of specified practices)" substitute "Registration under regulation 6 (registration of certain practices) in relation to the use of electrical equipment intended to produce x-rays for the purpose of research or the exposure of persons for medical treatment, and consent under regulation 7 (consent to carry out specified practices) in relation to the practices specified in regulation 7(1)(d), (e) and (f)".

The Health and Safety and Nuclear (Fees) Regulations 2016
16.   (1)   The Health and Safety and Nuclear (Fees) Regulations 2016 are amended as follows.

(2)   In regulation 2(1), for the definition of "the 1999 Regulations" substitute—
"the 2017 Regulations" means the Ionising Radiations Regulations 2017 [SI 2017/1075];

(3)   In the heading of regulation 8, for "the Ionising Radiation Regulations 1999" substitute "the Ionising Radiations Regulations 2017".

(4)   In regulation 8—

(a)   in paragraph (2)—
(i)   for "1(c)(i) or 1(d)(i)" substitute "1(d)(i) or 1(e)(i)";
(ii)   for "1999" substitute "2017";
(b)   after paragraph (2) insert—
"(2A) A fee is payable to the appropriate authority (within the relevant meaning given in the 2017 Regulations) on each application for registration or for a consent to carry out specified practices for

Schedule 9

the purposes of the 2017 Regulations."

(c) in paragraph (3), for "paragraph (1) or (2)" substitute "paragraph (1), (2) or (2A)";

(d) in paragraph (7), for "this regulation" substitute "paragraph (1), (2) or (4)";

(e) in paragraph (9), for "regulation 21(3)(e) of the 1999 Regulations" substitute "regulation 22(3)(e) of the 2017 Regulations".

(5) In Schedule 4, in relation to entry (a)—

(a) for "The 1999 Regulations" substitute "The 2017 Regulations";

(b) for "SI 1999/3232" substitute "SI 2017/1075".

(6) In the heading of Schedule 6, for "1999" substitute "2017".

(7) In Schedule 6, in column 1 of Table 1—

(a) for "regulation 35 of the 1999 Regulations", in both places in which it occurs, substitute "regulation 36 of the 2017 Regulations";

(b) in the entry for "Original type approval of apparatus"—
(i) for "paragraph 1(c)(i) or 1(d)(i) of Schedule 1 to the 1999 Regulations" substitute "paragraph 1(d)(i) or 1(e)(i) of Schedule 1 to the 2017 Regulations";
(ii) for "regulation 6" substitute "regulation 5";

(c) in the entry for "Amendment of an original approval of dosimetry services", in addition to the amendment made by sub-paragraph (a) above, for "paragraph 1(c)(i) or 1(d)(i)" substitute "paragraph 1(d)(i) or 1(e)(i)";

(d) after the entry referred in sub-paragraph (c) above, insert— "Application for registration or for consent to carry out a specified practice pursuant to regulations 6 and 7 of the 2017 Regulations."

(8) In Schedule 6, in column 2 of Table 1, in relation to the entry inserted by paragraph (7)(d), insert "£25".

(9) In Schedule 6, in column 1 of Table 2, for "regulation 35 of the 1999 Regulations", in both places in which it occurs, substitute "regulation 36 of the 2017 Regulations".

(10) In Schedule 6, in column 3 of Table 3, for "regulation 21(3)(e) of the 1999 Regulations" substitute "regulation 22(3)(e) of the 2017 Regulations".

The Environmental Permitting (England and Wales) Regulations 2016
17. In Part 5 of Schedule 23 to the Environmental Permitting (England and Wales) Regulations 2016, omit paragraph 7.

APPENDIX 1 Notice of approval

By virtue of section 16(4) of the Health and Safety at Work etc Act 1974, and with the consent of the Secretary of State for Work and Pensions, the Health and Safety Executive has on 2 October 2017 approved the Code of Practice entitled *Work with ionising radiation, Ionising Radiations Regulations 2017 Approved Code of Practice and guidance* (2018, L121).

The Code of Practice gives practical guidance on the Ionising Radiations Regulations 2017 (SI 2017/1075).

By virtue of section 16(5) of that Act and with the consent of the Secretary of State for Work and Pensions, the Health and Safety Executive has withdrawn its approval of the Code of Practice entitled *Work with ionising radiation, Ionising Radiations Regulations 1999 Approved Code of Practice and guidance* (1999, L121) which shall cease to have effect on 1 January 2018.

This Code of Practice (2018, L121) comes into effect on 23 March 2018.

Signed

Secretary to the Board of the Health and Safety Executive
16 February 2018

APPENDIX 2 Estimating effective dose and equivalent dose from external radiation

Effective dose

Effective dose from external radiation is generally defined by the relationship:

$$E_{ext} = \sum_T W_T H_T = \sum_T W_T \sum_R W_R D_{T,R}$$

This represents the sum of the weighted equivalent doses in all the tissues and organs of the body from external radiation, where:

- $D_{T,R}$ is the absorbed dose averaged over the tissue or organ T, due to radiation R;
- H_T is the equivalent dose;
- W_R is the radiation weighting factor; and
- W_T is the tissue weighting factor for tissue or organ T.

The total effective dose, E, includes the 50-year committed effective dose from internal radiation, E_{int}, from inhaled or ingested radionuclides usually based on radionuclide-specific dose per unit intake values.

Equivalent dose

Equivalent dose (H_T) is the absorbed dose, in tissue or organ T weighted for the type and quality of radiation R. It is given by:

$$H_{T,R} = W_R D_{T,R}$$

where:

- $D_{T,R}$ is the absorbed dose averaged over the tissue or organ T, due to radiation R; and
- W_R is the radiation weighting factor.

When the radiation field is composed of types and energies with different values of W_R, the total equivalent dose, H_T is given by:

$$H_T = \sum_R W_R D_{T,R}$$

Estimating effective dose

Where the effective dose from external radiation is estimated from personal monitoring, the operational dose quantity *personal dose equivalent Hp(d)* is normally used to demonstrate compliance with dose limits.

Personal dose equivalent is the dose in soft tissues at an appropriate depth, *d* in mm, below a specified point in the body and is given in sieverts.

Where doses are to be estimated from area monitoring results, the relevant operational quantities are:

- ambient dose equivalent *H*(d)*; and
- directional dose equivalent *H' (d, Ω)*.

In these cases, *d* is the depth in mm under the surface of the International Commission on Radiation Units and Measurements (ICRU) sphere. For strongly penetrating radiation a depth of 10 mm is appropriate and for weakly penetrating radiation a depth of 0.07 mm for the skin and 3 mm for the eye is recommended. Ω is the angle of incidence.

Ambient dose equivalent is the dose equivalent at a point in a radiation field that would be produced by the corresponding expanded and aligned field in the ICRU sphere at a depth, *d*, on the radius opposing the direction of the aligned field and is given in sieverts.

Directional dose equivalent is the dose equivalent at a point in a radiation field that would be produced by the corresponding expanded field, in the ICRU sphere at a depth, *d*, on a radius in a specified direction, Ω, and is given in sieverts.

The ICRU *sphere* is a body introduced by the ICRU to approximate the human body as regards energy absorption from ionising radiation; it consists of a 30 cm diameter tissue equivalent sphere with a density of $1\,g\,cm^{-3}$ and a mass composition of 76.2% oxygen, 11.1% carbon, 10.1% hydrogen and 2.6% nitrogen.

An *expanded field* is a field derived from the actual field, where the fluence and its directional and energy distributions have the same values throughout the volume of interest as in the actual field at the point of reference.

An *expanded and aligned* field is a radiation field in which the fluence and its directional and energy distribution are the same as in the expanded field but the fluence is unidirectional.

The fluence, Φ is the quotient of *dN* by *da*, where *dN* is the number of particles which enter a sphere of cross-sectional area *da*.

The *quality factor, Q*, is a function of linear energy transfer (*L*) in water and is used to weight the absorbed dose at a point in such a way as to take into account the quality of a radiation in a specified tissue or organ.

Unrestricted linear energy transfer (*L*) is the quotient of *dE* by *dl*, where *dE* is the mean energy lost by a particle of energy *E* in traversing a distance *dl* in water.

The relationship between the quality factor *Q(L)* and the unrestricted energy transfer *L* is as follows.

Unrestricted linear energy transfer, *L*, in water ($keV\,\mu m^{-1}$)	*Q(L)*
<10	1
10–100	0.32L–2.2
>100	$300\,\sqrt{L}$

Tissue weighting factors

Appropriate values of tissue weighting factor (W_T) to be used to weight the equivalent dose in a tissue or organ (T), where necessary.

Tissue or organ	Tissue weighting factor
Bone marrow (red)	0.12
Colon*	0.12
Lung	0.12
Stomach	0.12
Breast	0.12
Remainder tissues	0.12 (†)(††)
Gonads	0.08
Bladder	0.04
Oesophagus	0.04
Liver	0.04
Thyroid	0.04
Bone surface	0.01
Brain	0.01
Salivary glands	0.01
Skin	0.01

*   Dose to the colon is taken to be the mass weighted average dose to the upper and lower large intestines.

†   For the purposes of calculation, the remainder is composed of the following additional tissues and organs: adrenals, extrathoracic (ET) region, gall bladder, heart, kidneys, lymphatic nodes, muscle, oral mucosa, pancreas, prostate, small intestine, spleen, thymus, uterus/cervix.

††   The equivalent dose to the remainder tissues is normally calculated as the mass-weighted mean dose to the organs and tissues listed above. In the exceptional case in which the most highly irradiated remainder tissue or organ receives the highest equivalent dose of all organs, a weighting factor of 0.025 (half of remainder) is applied to that tissue or organ and 0.025 to the mass weighted equivalent dose in the rest of the remainder tissues and organs.

Radiation weighting factors

Where required for a direct estimate of E, values of radiation weighting factor, W_R, depend on the type and quality of the external radiation field.

Type and energy range	Radiation weighting factor, W_R
Photons, all energies	1
Electrons and muons, all energies	1
Neutrons	A continuous function of neutron energy (see below)
Protons and charged pions	2
Alpha particles, fission fragments, heavy ions	20

In calculations involving neutrons a continuous function of neutron energy should be used:

$$W_R = \begin{cases} 2.5 + 18.2e^{-[\ln(E)]^2/6}, & E < 1 \text{ MeV} \\ 5.0 + 17.0e^{-[\ln(2E)]^2/6}, & 1 \text{ MeV} \le E \le 50 \text{ MeV} \\ 2.5 + 3.25e^{-[\ln(0.04E)]^2/6}, & E > 50 \text{ MeV} \end{cases}$$

where E is the neutron energy. HSE may authorise the use of equivalent methods for estimating E.

APPENDIX 3 Abbreviations

ACOP   Approved Code of Practice

ALARP   as low as is reasonably practicable

HSE   Health and Safety Executive

HSWA   Health and Safety at Work etc Act 1974

NORM   Naturally occurring radioactive materials

ONR   Office for Nuclear Regulation

PPE   personal protective equipment

RPA   radiation protection adviser

RPE   respiratory protective equipment

RPS   radiation protection supervisor

SFARP   so far as reasonably practicable

References

1. The Ionising Radiations Regulations 2017 SI 2017/1075
   www.legislation.gov.uk/uksi/2017/1075/contents/made

2. Council Directive 2013/59 Euratom of 5 December 2013, laying down basic
   safety standards for protection against the dangers arising from exposure to
   ionising radiation. *Official Journal of the European Union L13*

3. Health and Safety at Work etc Act 1974 The Stationery Office
   ISBN 978 010 543774 1 www.legislation.gov.uk/ukpga/1974/37/contents

4. Safety Representatives and Safety Committees Regulations 1977
   The Stationery Office www.legislation.gov.uk

5. *Consulting workers on health and safety. Safety Representatives and Safety
   Committees Regulations 1977 (as amended) and Health and Safety
   (Consultation with Employees) Regulations 1996 (as amended). Approved
   Codes of Practice and guidance* L146 (Second edition) HSE Books 2012
   ISBN 978 0 7176 6461 0 www.hse.gov.uk/pubns/books/L146.htm

6. Ionising Radiation (Medical Exposure) Regulations 2017 and guidance
   www.gov.uk/government/publications/the-ionising-radiation-medical-
   exposure-regulations-2017

7. *Safety signs and signals. The Health and Safety (Safety Signs and Signals)
   Regulations 1996. Guidance on Regulations* L64 (Third edition) HSE Books
   2015 ISBN 978 0 7176 6598 3 www.hse.gov.uk/pubns/books/l64.htm

8. Personal Protective Equipment at Work Regulations 1992 (as amended)
   The Stationery Office www.legislation.gov.uk

9. Personal Protective Equipment Regulations 2002
   The Stationery Office www.legislation.gov.uk

10. *Personal Protective Equipment at Work (Second edition). Personal Protective
    Equipment at Work Regulations 1992 (as amended). Guidance on Regulations*
    L25 (Third edition) HSE Books 2015 ISBN 978 0 7176 6597 6
    www.hse.gov.uk/pubns/books/L25.htm

11. Safety assessment principles for nuclear facilities http://www.onr.org.uk/saps

12. Employment Rights Act 1996
    The Stationery Office ISBN 978 0 10541896 2 www.legislation.gov.uk

13. *Working safely with ionising radiation: Guidelines for expectant or breastfeeding
    mothers* INDG334(rev1) HSE 2015 www.hse.gov.uk/pubns/indg334.htm and

New and expectant mothers who work: A brief guide to your health and safety
Leaflet INDG373(rev2) HSE 2013 www.hse.gov.uk/pubns/indg373.htm

14. Nuclear Installations Act 1965
     The Stationery Office ISBN 978 0 10 850216 3

15. *Respiratory protective equipment at work: A practical guide* HSG53 HSE
     Books http://www.hse.gov.uk/pubns/books/hsg53.ht

16. ICRP Publications 116/119
     www.icrp.org/publication.asp?id=ICRP%20Publication%20116
     www.icrp.org/publication.asp?id=ICRP%20Publication%20119

17. British Institute of Radiology 'Patients leaving hospital after administration of
     radioactive substances' BIR Publications

18. Environmental Permitting Regulations (England and Wales) 2016
     The Stationery Office ISBN 978 0 11 115247 8 www.legislation.gov.uk

19. Radioactive Substances Act 1993
     The Stationery Office ISBN 978 0 10 541293 9 www.legislation.gov.uk

20. HSE statement on radiation protection advisers available on HSE's website:
     www.hse.gov.uk/radiation/rpnews/statementrpa.htm

21. National Physical Laboratory measurement and good practice guide: *The
     examination, testing and calibration of portable radiation protection
     instruments*. Available from the National Physical Laboratory www.npl.co.uk

22. Data Protection Act 1998
     The Stationery Office ISBN 978 0 10542998 2 www.legislation.gov.uk

23. Central Index of Dose Records on HSE's website: www.hse.gov.uk/radiation

24. Special entries in dose records, information available on HSE's website:
     www.hse.gov.uk/radiation/ionising/doses/

25. ISO 9978 *Radiation Protection sealed radioactive sources*
     www.iso.org/standard/17886.html

26. Carriage of Dangerous Goods and Use of Transportable Pressure Equipment
     Regulations 2009 ONR www.onr.org.uk/transport/guidance.htm

27. HSE's RIDDOR (Reporting of Injuries, Diseases and Dangerous Occurrences)
     website: www.hse.gov.uk/riddor/

28. Medical Devices Regulations 2002 (as amended)
     The Stationery Office www.legislation.gov.uk

Further information

For information about health and safety visit https://books.hse.gov.uk or http://www.hse.gov.uk.

You can view HSE guidance online and order priced publications from the website. HSE priced publications are also available from bookshops.

To report inconsistencies or inaccuracies in this guidance email: commissioning@wlt.com.

British Standards can be obtained in PDF or hard copy formats from BSI: http://shop.bsigroup.com or by contacting BSI Customer Services for hard copies only. Tel: 0846 086 9001 email: cservices@bsigroup.com.

The Stationery Office publications are available from The Stationery Office, PO Box 29, Norwich NR3 1GN Tel: 0333 202 5070 Fax: 0333 202 5080. E-mail:customer.services@tso.co.uk Website: www.tso.co.uk. They are also available from bookshops.

Statutory Instruments can be viewed free of charge at www.legislation.gov.uk where you can also search for changes to legislation.